FOCUS OF QUALITY ON QAULIATY ASSURANCE OF A MINI CUSTOM POWER PARK

TANIA

TABLE OF CONTENTS

CHAPTER 1

INTRODUCTION

1.1. General

Nowadays Power Quality is a major concern in electric power sector as it has become an important measure of service reliability to both consumers and utility. The introduction of power electronic devices in domestic and industrial sectors, increased use of nonlinear loads, integration of renewable energy sources with grid, communication and other solid state devices in smart grid etc. are creating new power quality issues which need to be addressed before the situation goes worse. Power Quality is good for an end user, if voltage and current are of rated magnitude and frequency and sinusoidal in wave shape. Power quality can be measured in terms of parameters like power frequency, voltage variation, harmonics, flicker, transients, waveform distortion etc. There are natural (lightning, equipment failure, faults) as well as deliberate reasons (use of power electronic device based loads) for poor power quality. Power quality issues adversely affect both utility and consumers. Some of the effects of poor power quality are power failures, breakdowns of equipment or malfunctions of devices, reduction of useful life due to overheating of machines, sensitive equipment's damage, interference in electronic communication lines, increased losses in the distribution system etc.

In the year 2012, India witnessed a large-scale blackout which affected almost 8 to 10 major states of the northern part of India. Depleted transmission network, stretching the generating units to peak load for long durations, underutilized hydro power station, poor power frequency profile, large reactive power consumption etc. are some of the reasons of the blackout. But the main issue was lack of regulation among State Level Load Dispatch Centers. Priority-based load shedding was not possible since the domestic loads, commercial loads and essential services like hospitals and industries are connected to common feeders. System was not designed to restore power to essential services in a short span of time. Absence of a program to conserve energy and the absence of load shedding policies also got highlighted [1].

In 2015, Power Grid Corporation of India Ltd. has measured the Power-Quality indices at various places covering the entire country through POWERGRID substations (at different voltage levels) and done the analysis of Power-Quality in the Indian power system [2]. Fig.1.list outs locations critical in terms of different power quality indices.

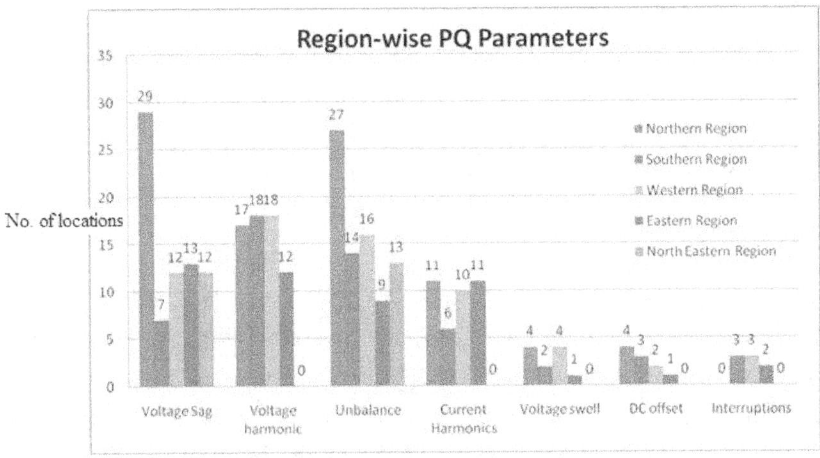

Fig.1.1. Overview of power quality parameters observed across the country [2]

From the chart it is clear that the total number of locations exceeding permissible limits across India are 73,65,79 in number out of 175 cities in the cases of voltage sag, voltage harmonics and unbalance respectively. These numbers indicate that even after the blackout that happened in 2012, the power quality issues have not properly been sorted out. Still in 2019, the condition is almost the same. However, the consumers are more aware in recent times due to the different types of penalties which have been levied on the consumers. These penalties are due to the various failures such as failure of equipment, wrong standards and interruptions.

With new initiatives like 'Make in India' proposed by the Indian Government, India started getting more attention from other countries in the field of setting up new commercial and manufacturing facilities. However, these companies do not have a good opinion about the quality of power in India and hence are reluctant to start their business initiatives. By considering all these factors, it can be stated that maintaining power quality has now become one of the major concerns of the Indian Power Sector. To ensure good quality of power, the assessment of power quality issues needs to be carried out first and then attempt on the mitigation techniques. Another factor to be considered is the deployment of the devices used for improving the power quality.

1.2. Motivation for Research

A survey which was conducted recently in the USA indicates that a loss of $45.7 billion per year is occurring to the industrial and digital business firms due to interruptions in electric power. The survey also says that around $160 billion loss is happening to all business sectors due to interruptions and another $25 billion due to power quality problems.

In India, power quality issues lead to up to 10% losses in the turnover of large companies. These include losses due to long and short interruptions, voltage dips and transients. The direct costs of downtime in India due to bad power quality (PQ) are to the tune of ₹20, 000 crores per annum. The indirect costs could be very high compared to this figure. Present day lifestyle expects a reliable and continuous supply of electric power. The solution for this requirement is to have smart grids with energy storage facility which will decrease the dependency on the supply from centralized power generation using conventional sources by introducing renewable energy sources such as wind, solar etc. However, this design increases the risk of power blackout due to increased dependency on highly volatile renewable energy sources. When a continuous power supply is available, power quality is the next requirement. Voltage quality is mainly affected by various events such as faults, harmonic disturbances and lightning strikes, etc. Other than this, equipment on the end-user side have less immunity to the voltage disturbances. Power quality issues will result in the malfunction or damage of the equipment. Since the use of electronic circuits to build devices from home appliances to large control units in industrial processes has increased multiple folds, the awareness of power quality is also increased.

In the modern grids, power comes from multiple sources, flows in multiple directions, and is more environment friendly. Since the power generators have varying parameters and due to the inclusion of renewable energy sources, voltage and power quality issues are prevalent. Some of the key Power Quality issues in modern grids are:

Power Quality issues introduced by new consumption models:

When smart grids are introduced to the environment, typical fossil-based fuels are replaced by electric / solar vehicles that demand charging stations for electric vehicles. This is a new type of service. Similarly, expanded high-speed railways is another new service introduced. These newly introduced services introduce power quality issues also.

Cross talk or interference between typical power line communication and electrical devices:

Smart grid concept is heavily dependent on inter communication of devices involved in the smart grid. It can be sensors, electrical devices, distribution components etc. It also needs to have communication between customers, distributed generators, and the grid operator. These communication modes can be wired or wireless. Power-line communication is an essential part due to its easy availability. But the power lines will introduce disturbances in the power system, resulting in poor power quality.

Distribution grid weakening:

The smart grids brought in an increased use of distributed generation and renewable energy sources, which reduces the usage of conventional power. Penetration of renewable energy into the grid has the adverse effect of creating power quality disturbances in large scale.

Immunity of device:

Voltage quality disturbances due to voltage dip trigger the simultaneous tripping of many distributed generators. Smart grid tries to keep a balance between generation and consumption. During this attempt, mass tripping of consumption will have consequences.

The most serious side effect of poor power quality is the damage caused to physical network. Loss due to data corruption, productivity reduction, loss due to increased downtime etc. are other consequences. Hence, monitoring power quality along with proper accounting, is the backbone of a fully implemented modern grid.

One has to understand the importance of growing consciousness of customers on power quality and rising voice of demand for high quality power. Despite power quality providing an opportunity for boosting their profits, power distribution companies are yet to embrace it. Since most of the end users have sensitive loads, the concept of user specific power distribution is a good idea and it is the motivation for Custom Power Parks (CPP). In this mechanism, the end users are getting an opportunity to demand the kind of power which they want because at the end, it is the same end users who pay for the in-efficiencies in power system network due to poor power quality. The custom power park uses mainly two categories of devices; network reconfigurable devices and compensating devices. Compensating devices include Series Active Filter, Shunt Active Filter and Unified Power Quality Conditioner. Sensitive loads are effected by sags and swells, harmonics and these issues reduce the power quality. In such situations, Series Active Filters are the best solutions. Shunt Active Filter performs the functions like balancing of source current, power factor correction, and harmonic mitigation. Supplying or accepting reactive power in the distribution system maintains bus voltage at the required level and Shunt Active Filters can be used to perform this. Unified Power Quality Conditioner performs the functions of both series and Shunt Active Filters. With these compensating devices almost all the issues related with voltage, current and frequency can be sorted out.

The network reconfigurable type devices consist of solid state transfer switch, solid state current limiter and solid state breaker. These devices employ advanced power electronic devices for fast acting operation. The role of these devices is mainly to maintain the reliability of the supply. In this way the two main issues; maintain reliability and power quality improvement, faced by modern grid or companies in the deregulated structure of the power system can be resolved by building a power park with custom power devices and controlling its operation by an automated and centralized controller. With this, it could categorize the loads into different grades according to the service requirements and sell to the end users. In modern times, in the consumer sector there is an ever increasing need of high quality power and increased expectations on power quality. Due to this custom power parks are getting more popularity. Low power quality level brings in heavy losses in the production sector. Thus, the single step answer to the modern day customers' needs is Custom Power Parks (CPP).

To understand the context in which a CPP fits well, consider a hospital where there can be several critical devices which are used in patient monitoring and treatment. Let us consider an artificial life support system in a hospital. Typically, the core of such a system will be a computer for the centralized data collection and processing. Power quality issues may cause damages to the microprocessor unit of the computer system and this may lead to malfunctioning or stoppage of the microprocessor unit. This will lead to the stoppage of the services provided by the Artificial Life support system. This is a complete failure of the system and cannot be allowed to occur. In such scenarios, Custom Power park concept in which end user gets specific quality power with improved reliability for different sensitive loads is the perfect solution.

The research work mainly focuses on developing the Custom Power Park in the Indian context. The primary targeted area in the implementation of CPP is the power quality improvement. The work involves development of the overall controller, which will do centralized control operation of the entire devices connected in the custom power park. By this mechanism, the customer need not have to install the custom power devices or compensation devices at his own premises, instead he can demand the quality of power which he wants to consume from the service provider. This thesis also focuses on developing a novel controller for the series active filter, which is one of the most crucial elements in the custom power park. The work does not involve the financial aspects of the CPP such as the tariff levied to end users, buying cost per unit of the power from various power generators etc. These aspects are left to the implementation authorities which will be involved in the deployment of the Custom Power Park in India.

1.3. State of the Art

Power quality issues came into picture since the inception of electrical power. The increased use of solid state controllers, digital controllers, power electronic based loads, variable speed drives and so on, has increased the magnitude of power quality issues. Now a days the consumers are more aware about these disturbances caused by the power quality issues to their equipment and are concerned about the direct and indirect penalty enforced on them. The deregulated structure and distributed generation also pollute the power network. In view of these issues, there is a remarkable growth in the development of mitigation techniques thereby maintaining the customer's equipment at improved power quality level. Passive filtering and Active filtering are now matured technology in power quality improvement. They are still widely used because of their advantages. Passive filters are simple, economic and easily realizable. Since the year 1984, different topologies of active filters have been proposed. All current related harmonic issues can be sorted out by the Shunt Active Filter. Since 1971 many configurations of active series filter have been proposed. Voltage related power quality issues can be eliminated by the use of active series filters. The major advancements in the case of the active filter are the use of fast acting, self-commutating solid state devices, different analog and digital controllers, and different types of sensors etc which help to improve its performance. Optimizing the control techniques is also the new research trend. Hybrid filters and Improved Power Quality Converters (IPQC)are also in use for the past decades. The two main blackouts that happened, in USA in 2003 and in Northern India in 2012, were the cumulative results of certain power quality issues. Even though now it is more or less fully interconnected networks, the coordination among different devices in smart grid is not proper. So, an overall coordinated control scheme is the main requirement in today's power system. In America, Japan and Canada, different CPDs were installed in 1990's itself. But the concept of CPP and its coordinated controller is still in infancy. Just like FACTS technology improves the reliability and quality of transmission, the custom power park improves the reliability and quality of power service at the distribution side. End users with sensitive loads require high quality power without interruption and thus Custom Power Parks with coordinated control is the need of the hour.

In Indian Scenario where the concept of smart grid itself is in initial stage of deployment, there is a huge scope existing for research in various aspects of smart grid. Considering the different issues faced like black outs, it is evident that a Custom Power Park with proper control is essential. The proposed scheme is the design and development of a low cost custom power park with an efficient controller which focuses on power quality improvements and does not consider the financial aspects like user tariffs and cost of power to be provided to power generation companies.

1.4. Scope of the work

This research work aims to design and develop a Custom Power Park which provides customized power service with high quality, regulated, uninterrupted, steady and clean electric power service as per the customer demand with the help of Custom Power Devices (CPDs). It includes the implementation of a novel and efficient controller for one of the CPDs, a Series Active Filter. The design, simulation, prototype development and testing and validation of the system is done here under varying conditions. The comparison of proposed controller with already existing controllers is also carried out to prove the effectiveness of the proposed controller. The prototype of CPP is developed in hardware. The Hybrid Solid State Transfer Switch (SSTS) is developed and with proper coordinated controller, the fast switchover between preferred and alternate feeders to the CPP is made possible. Smart sensing and communication is deployed across the Custom Power Plant and all the control actions are performed using this, thus making the proposed model 'Smart'. The readings from the sensors are communicated to the control unit via wifi modules. Control unit decides the switching of various CPDs & loads and communicates back necessary information via wifi modules. The work mainly prioritizes the power quality improvement schemes in the CPP. Since in Indian context, the deployment of CPP is typically carried out by the government or government supported agencies, the financial aspects of the CPP is not looked into much. These aspects include the cost involved in buying power from various players in the power generation domain, parameters for finalizing the set of power generation companies and the tariff for various grades of power.

1.4.1. Objectives

The main objectives include,

i. Design and modeling of the proposed controller for series active filter

ii. Comparison of the performance analysis of the proposed controller with already existing control algorithms such as Instantaneous Reactive Power Theory (IRPT) and Synchronous Reference Frame (SRF).

iii. Hardware implementation of the proposed controller for the series active filter and analysis of the results

iv. Simulation of the proposed Custom Power Park (CPP) and analysis of the performance of overall controller

v. Performance analysis of different Custom Power Devices (CPDs)

vi. Hardware implementation of the prototype of proposed Custom Power Park

1.4.2. Methodology/ Work plan

Methodology to carry out the investigation includes

i. Design, simulation and analysis of proposed controller for Series Active Filter using MATLAB/Simulink platform

ii. Simulation of Instantaneous Reactive Power Theory based and Synchronous Reference Frame based control algorithms in MATLAB/Simulink platform and performance comparison with proposed control algorithm.

iii. Experimental analysis of the proposed controller for Series Active Filter.

iv. Design and development of a coordinated controller for CPP in MATLAB/Simulink platform as well as performance analysis by building the hardware set up.

v. Implementation and testing of hardware prototype of Solid State Transfer Switch (SSTS) with faster response time.

vi. Simulation of CPP and performance analysis of different CPDs

vii. Experimental set up of CPP for single phase system.

viii. Development of communication module.

1.5. Organization of the report

The thesis starts with the Introduction to the concept of Custom Power Parks, their need, the scope of this thesis work with defined objectives in Chapter 1. Chapter 2 gives a brief survey report of the relevant literature related to the research. In Chapter 3, the various power quality issues, improvement techniques and standards are discussed.

Chapter 4 presents the layout of the proposed Custom Power Park. Chapter 5 gives the detailed description of the newly proposed controller for a Series Active Filter, its simulation analysis, hardware implementation and test results under various conditions. Chapter 6 contains the simulation analysis of various custom power devices, explanation   about hardware prototype of SSTS and CPP setup. It also presents the working of overall coordinated controller with communication module integrated with the CPP setup.

The report comes to an end with conclusion of the work in Chapter 7 followed by references and appendix.

CHAPTER 2

LITERATURE SURVEY

2.1. Introduction

The main focus of the research work is Power Quality and various means to improve the power quality. This chapter provides the details of the thorough literature survey carried out on different aspects of the topic of this research work. The various aspects looked into are power quality issues, causes and effects, PQ improvement techniques, Custom Power Park concept and Custom Power Devices, Series Active Filter control algorithms and Shunt Active Filter control algorithms.

2.2 Power Quality Issues, Causes, Effects and Improvement Techniques

With the advancement of power electronics in industry, power quality issues have caused significant impact on the electrical grid for the past two decades. It has introduced pollution of power in generation, transmission and distribution systems. The main PQ issues are voltage variations, transients, frequency variations, brown out, black out, harmonics, interruption and flicker [3]. Voltage variations can be long duration and short duration voltage variations based on the duration it exists. The proliferation of voltage sensitive, semiconductor based or computer based or adjustable speed drives into the field of power system and industries has made the necessity of maintaining the voltages exactly within the limits more stringent. Even if the voltage sag is for small duration it causes significant effects in the performance of these devices. Voltage transients cause computers and other equipment to experience flickering, equipment shut off and loss of data. Voltage variations are caused by switching on/off of capacitor banks/transformer, faults, network loading effects, control system imperfections on voltage regulator or transformer tap setting etc [4]. The main cause of frequency variation is erratic operation of generators or unstable frequency power sources which cause program failure, data losses, equipment lock up or complete shutdown. When there is peak load demand especially in summer, the utilities may fail to meet the requirement, must lower the voltage to limit maximum power. This causes brown out where as black out is caused by power distribution failure or utility failure. The main reason for harmonics are nonlinear loads such as power electronic equipment, Switch Mode Power Supplies (SMPS), Adjustable speed drives (ASDs), arc furnaces, welding machines etc [5]. Power quality indices like Total Harmonic Distortion(THD), Total Demand Distortion(TDD), Harmonic Factor(HF), I.T product, C message Index etc are used to measure the severity of different PQ issues [6]. The power quality issues cause adverse effect on customer as well as utility side. Some of the effects are increased heating in cables and other equipment, neutral wire over loading, errors in measurements of meters, maloperation of protective devices, damage of sensitive equipment, loss of information and maloperation of data processing equipment, increased probability of occurrence of resonance, aging of capacitor banks and machines etc.

There are some possible solutions to mitigate these PQ problems. Surge suppressors, noise filters, voltage regulators, UPS, power conditioners and proper grounding and proper wiring may reduce the PQ issues to a certain extent. Another development is the invention of harmonic filters. The hierarchy of improvement techniques is in such a way that initially fixed capacitors and reactors and synchronous condenser came into picture. Then in 1970's FACTS devices (both variable impedance type and switched converter type) became the necessary technique. Later on, introduction of passive, active and hybrid filters for mitigating power quality issues became a reality [7].

Passive filters offer resonance at tuned frequencies thereby eliminating particular frequency components. But the main drawback of passive filter is its size and in certain situations it may act as load creating burden on the utility, especially at fundamental frequency. An alternative to this is to use active filters. Based on the application different active filter configurations can be used. Shunt Active Filter can be operated in current control and voltage regulator mode so as to provide current harmonic compensation, reactive power compensation, flicker elimination, compensating unbalanced currents and maintaining rated voltage at bus terminals respectively [8-9]. The Series Active Filter injects voltage in such a way that it will be restore the terminal voltage to rated voltage of fundamental frequency at the load bus. The main functions of Series Active Filter are voltage harmonic compensation, compensation of sag and swell etc. Active filters have better filtering performance than passive filters [10]. The only drawback is the cost. To overcome this, combination of active and passive filters can be used. Different hybrid filter configurations have been proposed in literature [11-13]. The shunt active power filters are mainly suitable for compensating harmonics in the current fed type harmonic producing loads. Unified Power Quality Conditioner (UPQC) is a hybrid filter in which shunt and Series Active Filters are connected back to back to perform the functions of both filters. UPQC is basically a single solution to mitigate voltage and current related problems. These can be classified in many ways, such as supply system based, topology based, converter based and method of control based classifications. Another option is the use of Improved Power Quality Converters (IPQCs) [14-15]. The main demerits like current harmonics, poor power factor and slow varying rippled dc output, low efficiency etc of conventional diode bridge rectifier can be overcome with the invention of IPQCs. By using power quality improvement techniques, the PQ issues can be bringing back to the limits specified by IEEE standards [16]. Several standards have been developed by different countries on the permissible levels of deviations of electrical quantities. IEEE standard 519-1192, IEC 61000 3-2, EN 50160 etc are examples of different standards on various issues of power quality [17].

2.3 Power Grids and Smart Grids

In most of the countries, the power generation and distribution is a process which involves multiples parties. There will be several power generation sources and companies who distribute the power to the end users. For an efficient power generation and distribution, these entities are connected together to form a power grid. Power grid basically consists of series of two components, loads and generators. Generators are the sources of power to the grid whereas loads or consumers are the one which uses the power from the grid. This system of inter-connected loads and generators is called grid [18].

Developments in the area of wireless communication and sensor networks, brought changes in the working of power grids also. The use of sensors to identify the demand for power in various parts of the grid and redesigning the distribution structure according to the dynamic requirement became more effective and such a system is referred to as smart grids [19-20].

The increased interest towards renewable energy has motivated the researchers to put effort in identifying schemes which enable the integration of renewable energy to smart grids. This will bring the end user a lot more stable power supply which is cost effective and environment friendly. At the same time, the integration of renewable energy to smart grids brings new challenges due to the behavior of renewable energy mainly variability and uncertainty. These factors introduce a new range of power quality issues and that need to be addressed [21].

Since all the power generators and distributors are interconnected, any mismatch in load or failures lead to complete blackout of the grid. In 2012, North India faced a complete blackout as the northern region grid failed. Even though the restoration process took place within few hours, many critical areas like hospitals and high security data centers faced huge damages and economic loss [1].

In addition to the chances of blackouts, power quality issues also introduced new challenges to the distributors as the end users started expecting high quality power for their regular uses [22-23]. A user may demand various levels of quality for the power he/she draws and is ready to pay accordingly. Current smart grid concept does not have the ability to deliver varying power quality levels to users based on demand. A solution to this scenario is the Custom Power Parks.

2.4 Concept of Custom Power

The consumers and utilities are now more concerned about the quality of power. The main reason behind that is, the loads which are based on embedded controllers and power electronic devices which are more sensitive to power quality variations. Consumers have increased awareness of power quality issues and they seek the utility to deliver high quality power. Another reason is that since most of the things are interconnected, the consequences of failure of one component will be more. Distributed Generation (DG) and globalization of the industries have increased the power quality problems. In many suburban regions there is no tightly coordinated control of the power from generation to end user. The consumers are not ready to install expensive compensating devices and utility has its own limitations. The common solution to address this problem is the development of custom power plants/parks. In this, the customers are getting an opportunity to buy the kind of power which they want. The main concept behind the custom power park is that it buys grid power and by installing many custom power devices, improves the quality and sells it at higher cost [24-25].

The concepts and operating principles of a Custom Power Park (CPP) is explained in detail by **Dr. Narain G. Hingorani** et all. This concept opens a host of new revenue opportunities for the forward-looking utilities. According to this the loads are categorized into three different grades based on the quality, Grade A, Grade AA and Grade AAA. Grade AAA has highest quality power [26].

Grade A category gets the improved quality power with the services of Shunt Active Filter and SSTS. Grade AA is superior to Grade A with an extra service of Diesel Generator. Grade AAA is the best quality of power as it gets the services of Series Active Filter as extra compared to Grade

AA. The customers can select the combination of these powers so that wherever critical loads are there, grade AAA power can be chosen and for non-critical loads, AA or A category can be selected. In this way cost optimization can be done.

2.5 Custom Power Devices

The Custom Power Devices (CPDs) are mainly two types, network reconfigurable type devices and compensating devices. The network reconfigurable type devices include SSTS, SSCL and SSB. All these devices are based on solid state devices and hence are very fast acting type with response time less than 5ms [27].

Compensating type CPDs include Shunt Active Filters and Series Active Filters. With proper control algorithm, Shunt and Series Active Filters are used to compensate the current and voltage related issues respectively. Combination of these also can be used. The services of Shunt Active Filter are provided to all grades of loads whereas the services of Series Active Filter are provided to only grade AAA loads which are most sensitive types loads which cannot even withstand very small changes in voltage itself [28-29].

There are various voltage detection schemes suggested in literature. Depending on the system condition suitable detection scheme can be selected. Some papers detail about the control schemes for the devices in the park to achieve the necessary compensation [30-31].

Static Transfer Switch which helps in switching between feeders when the voltage at a particular feeder falls below certain limit, is an important device in the park. Thyristor based or Gate Turn off (GTO) based static transfer switch can be used according to the application. The GTO based static transfer switch is preferred to achieve fast control of the fault current. The different topologies for SSTS is also proposed in different literature [32-34].

When the end users of a smart grid are considered, there are many areas which has highly sensitive loads. For example, a modern hospital has remote patient monitoring systems, fully autonomous artificial life support systems etc. Most of these systems are controlled by large computational clusters and these computational clusters are highly sensitive loads. Any small variation in power introduces performance issues and in extreme cases, it stops the functionalities also. It is therefore essential to provide quality power without having any power quality issues and this can be done with custom power parks equipped with power quality improvement techniques.

Among the CPDs the most important device is Series Active Filter/DVR which is mainly used to protect essential and sensitive loads from voltage disturbance [35-39]. The focus of the proposed scheme is mainly on designing an appropriate control strategy for the Series Active Filter for the CPP. A detailed literature review on Series Active Filters show that there is no research work carried out towards the development of an efficient controller for Series Active Filter in a custom power park setup.

2.6 Control Algorithms for Series Active Filter Control

The main objective of the control algorithm for Series Active Filters is to estimate the reference filter voltage to carry out necessary compensation. There are time domain as well as frequency domain based control strategies. For Series Active Filters the most commonly used algorithms are Instantaneous Reactive Power Theory [40-42] and Synchronous Reference Frame based algorithms [43-45]. The other algorithms that have been proposed are also slight variations of these two which includes either Clarke's or Park's transformation [46-50]. Park's or Clarke's transform involves converting three phase to two phase quantities by mapping to different reference frames. These involves more computation time and hence large memory. In IRPT first the signals are mapped into $\alpha\beta$ axis and instantaneous powers are computed. From the powers reference signals can be estimated. One cycle control and generating the reference by extracting positive and negative sequence components of voltages has also been in use [51-52]. To extract the positive and the negative sequence components from the fundamental and the harmonic waves so as to analyse and adjust each component separately, a multiple DQ transformation with separation modules has been used. In one cycle control method different switching strategy is used to generate pulses to the DVR instead of the conventional PWM techniques. Another method proposed to generate signals are based on symmetrical components [53]. The main drawback with all these control methods is the complex computation which increases the time as well as the cost. Nonlinear control techniques and advanced version of feed forward control strategies are also used to generate reference signals. [54-55]. Also, these controllers will not perform well under distorted source voltage condition. Thus, an efficient controller requirement is there which performs as expected under all conditions of system variations.

2.7 Control Algorithms for Shunt Active Filter Control

The heart of active filters is the control algorithm used to generate the reference signals. There are different control algorithms like Instantaneous Reactive Power Theory [56], Synchronous Detection [57], DC bus voltage algorithm [58], Synchronous Reference Frame [59] and IcosΦ algorithm [60], proposed by different people for the Shunt Active Filter. IRPT and synchronous reference based algorithms are based on clarke's and Park's transforms in which three phase time varying quantities are mapped into stationary and rotating reference frames respectively. In Synchronous Detection Control Algorithm three phase power is computed and from this DC component is extracted and distributed to each phase. The main principle behind DC bus voltage algorithm is that the voltage across the DC link capacitor indicates the information about real power flow and hence power imbalance between source and the load. According to IcosΦ control algorithm the source has to supply only real part of fundamental component of current, remaining will be taken care by the filter. Modified versions of IRPT and SRF algorithms are also been proposed in literature [61-62]. Among these the most efficient one is IcosΦ in terms of response time and filtering performance [60]. Few papers on fuzzy controller also have been referred to get an idea about fuzzy logic based controller for shunt and Series Active Filters [63-67]. When the input is partial/biased/incomplete and if the system cannot be represented using a mathematical model then fuzzy system will give the correct results. Artificial Neural Network based controllers and combination of different soft computing techniques based controllers are also proposed in literature [68-69].

2.8 Wireless Communication Schemes for Smart Grid

The rise of smart grid concept has introduced the need for communication among the various end points of the smart grid. Mainly there are two pairs of communication is required. The first one is between consumer premises and utility billing departments and the second one is between utility power control distribution centre and power generation facility. Some researchers have proposed Power Line Communication in these segments, but the major challenge is the heavy installation cost. This has opened the path for Wireless Communication Technology (WCT) to be used as the communication scheme in smart grids [70].

The massive growth in communication technologies have opened up a large research area for scientists to work on integrating wireless communication schemes to smart grids to make it more fast, effective and efficient in terms of decision making. Various wireless communication schemes such as ZigBee, WiFi etc can be exploited to extend the communication facility in smart grids and make it more distributed in nature [71-72]. Literature review indicates that, most of the time, low cost communication schemes such as Bluetooth or ZigBee are used as standards for wireless communication as they are already experimented [73-74]. In the proposed system, an effort is made to design a WiFi based communication scheme which has better coverage in terms of geographical area and less noise disturbance compared to traditional low-cost schemes [75-76].

An approach which is somewhat similar to the proposed Custom Power Park concept is implemented in Columbus, Ohio, USA. The entity is named as Premium Power Park (PPP). The main objective of PPP is to provide premium power to end users. To support this requirement, additional objectives are identified such as to identify cost effective devices to mitigate PQ problems, design a control scheme to coordinate between various devices. The park also monitors the data and compares it with the benchmarked data for performance analysis. The premium Power Park concept is deployed by two companies American Electric Power (AEP) and S&C Electric Company (SEC). The whole implementation and deployment of PPP is carried out in 3 phases. To evaluate the effectiveness of the PPP, two customers were identified based on the different power quality and reliability needs. The 3 phases of the PPP are:

- Phase 1 in which the decision making on various devices, types of customers and site for the implementation are considered.
- Phase 2 in which the actual implementation of the park is carried out.
- Phase 3 in which based on the collected data, performance analysis is carried out where both economic and technical aspects are analyzed.

In phase 1, various Power Quality devices identified are DVR, FASTRAN, Intellivar for the PPP implementation. The control signal used for DVR is based on abc to dq transformation which is comparatively complex in nature. Also FASTRAN is a high speed mechanical transfer switch which consumes more than a cycle time (25 ms) for switchover. Intellivar is an Advanced Solid State Static Var Compensator (ASVC) type controller(TSC+TCR) for eliminating voltage sags and voltage flickers and reactive power compensation [77-79]. From these configurations it is observed that the selection is not really an effective one and in consecutive phases of the PPP implementation, few modifications are carried out. Since dynamic compensation is not possible using Intellivar, and some issues which surfaced during the operation of PPP, Intellivar is removed from that system.

In the proposed system in this research work, in place of Intellivar, Shunt Active Filters are used which can provide dynamic compensation. Also the control algorithm selected for the DVR in the PPP implementation being comparatively complex in nature, a novel and effective control algorithm is suggested in the proposed system. In addition to these modifications, the proposal of a coordinated controller along with an effective communication scheme among the various entities of CPP is another major contribution of this work.

2.9 Summary of Literature Review Done

Papers related to power quality issues, their effects and improvement techniques are examined. A detailed study of the various custom power devices has been carried out. The need and requirements of custom power park is analyzed thoroughly. During the literature review, commonly used control algorithms for Series Active Filters are taken for detailed study. Papers on sag detection scheme and different topologies of SSTS are also referred. Some researchers have explained about the coordinated control scheme to distribute the power among different grades of loads.

CHAPTER 3
POWER QUALITY

3.1. Introduction

Nowadays processed power is used by electrical equipment deployed in industries in order to reduce power consumption, mechanical maintenance and cost of production. For this processed power, power converters are used which draw non-sinusoidal current contributing to degradation of power quality in the systems. Power converters generally consist of rectifiers at front end which inject harmonic current due to their nonlinear nature. These harmonics causes many problems in the power systems such as transformer overheating, harmonics, resonance in power systems, increased losses, malfunction of protection devices etc. Several techniques are proposed for the improvement of power quality. This chapter discusses about the various power quality issues, sources, effects of poor power quality and standards. It also discusses about the various power quality improvement techniques.

3.2. PQ Issues, Sources, Effects and Standards

Power Quality: The term power quality refers to any disturbance or irregularities in voltage, current or frequency which creates damage, malfunction or affects normal functioning of equipment used in the system or even cause system failure. The term power quality also describes the ability to properly function a load with the help of electric power.

The main power quality issues, sources and effects are given as:

Transients: Sudden change in electrical system parameters like voltage /current when it changes from one steady state operating condition to another is termed as transients. Usually the duration will be less than one cycle. Depending on the magnitude and duration it can be further categorized as impulse and oscillatory transients. Faults on feeders, lightning strikes, load changes, energization of transformers etc. can cause transients.

Long Duration Voltage Variations: Variations in supply voltage rms for a duration greater than one minute are considered as long duration voltage variations. This can be further classified as over voltages, under voltages and sustained interruptions.

Over voltage: 10% or more increase in fundamental frequency voltage for a duration greater than one minute. The main causes for this are switching OFF of large loads and switching ON of capacitor banks.

Under voltage: 10% or more decrease in fundamental frequency voltage for a duration greater than one minute. It results by the events which are reverse of the events causes overvoltage.

Sustained Interruptions: Discontinuity in electrical supply or decrease in voltage less than 0.1pu for duration more than one minute. Failure of equipment in the power system network, objects coming into contact with the lines or poles, fire, human error, improper coordination or protection devices failure can cause sustained interruptions.

As a result of interruption, functioning of electrical equipments freezes.

Short Duration Voltage Variations: Variations in rms value of supply voltage for a duration less than one minute are considered as short duration voltage variations. This is further classified as voltage sag, swell and interruptions.

Voltage Swell: Increase of 10% or more in the supply voltage, at the fundamental frequency with duration of more than one cycle and typically less than a few seconds. Single line to ground faults, Ferro resonant transformers may cause voltage swell.

Voltage Sag: A decrease of the normal voltage level of the nominal supply voltage at the fundamental frequency, for a duration less than 1 minute. These are also mainly caused by faults in feeders and connection of heavy loads and start-up of large motors.

Interruption: Decrease in the supply voltage less than 0.1pu for a duration less than 1 minute. It is mainly caused by the temporary faults in power system.

Both long and short duration voltage variations can cause malfunction of equipment used in microprocessor-based control systems, tripping of protective devices, damage and increased losses in electrical equipments etc.

Waveform Distortion: Distortion in the voltage waveform can be termed as waveform distortion. This is also further categorized into five types.

DC offset: Presence of DC component in AC system which are results of improper grounding, half wave rectification and geomagnetic disturbance. This can cause saturation of transformer core and excessive heating.

Harmonics: Voltage or Current waveforms having frequencies that are integer multiples of power-system frequency. Nonlinear loads cause harmonics. This mainly affects power system performance via; increased losses and decreased efficiency of equipments and cables, over loading of the neutral conductors, nuisance tripping of relays and circuit breakers, erroneous reading in measuring devices, increased probability of occurrence of resonance and electromagnetic interference with communication system etc.

Inter harmonics: Voltages or currents which contain frequencies that are non-integer multiples of fundamental frequency, usually those are results of frequency conversion.

Notching: Periodic voltage distortion caused by momentary short circuit of two phases of power converters when current commutates from one phase to another.

Noise: Superimposition of high frequency signals on the waveform of the power-system frequency signal caused by improper grounding and electromagnetic interference. This creates disturbance to sensitive equipments and also causes loss of data in data processing.

Voltage Unbalance: When all the voltages in three phase system are not equal in magnitude or equally displaced in time, which mainly occurs due to single phasing. These are mainly harmful to three phase loads like induction motors.

Voltage Fluctuations: Random variations in system voltage caused by the variations in the currents of the loads like arc furnace, starting current in induction motors etc. The main perceptible effect is the flickering of lighting and screens, causes uneasiness in visual perception.

Power frequency Variations: Variations in supply frequency due to load changes. It affects the generator speed and reduces the life of blades of turbine shaft.

Standards:

Consumer's equipments are greatly affected by electromagnetic interference. So there is a need to put a limit on harmonic emissions from the loads so that control on electromagnetic environment can be achieved. These kind of problems can be addressed by standards developed by certain organizations. The organizations responsible for developing power quality standards in the United States include the following:

Institute of Electrical and Electronics Engineers (IEEE)

American National Standards Institute (ANSI)

National Institute of Standards and Technology (NIST)

National Fire Protection Association (NFPA)

National Electrical Manufacturers Association (NEMA)

Electric Power Research Institute (EPRI) and

Underwriters Laboratories (UL).

The primary organizations responsible for developing international power quality standards outside the United States are International Electro Technical Commission (IEC), Euronorms, and ESKOM etc.

Power Quality standards define the quality of supply by determining the margins in which frequency and voltage are allowed to vary by putting limits on the harmonic current and voltage distortion, voltage fluctuations and duration of interruption etc. Some of the standards developed for PQ and measurements are the EN 50160, the IEC 61000-4-30 and the IEEE Standard 1159. EN 50160 is used to specify the characteristics of voltage in terms of voltage level, frequency and symmetry of the network at PCC etc. IEC 61000 also one of the first standards developed to denote the characteristics of supply voltage waveform as obtained from measurements. IEC 61000-3-2 standard puts limits on harmonic currents emission. In that standards, the equipments are classified as class A, class B, class C and class D equipments and according to each class the maximum harmonic current limit is defined.

IEEE 519 standard recognizes the responsibility that users should not degrade the voltage quality by serving other loads which draws nonlinear current and also recognizes the responsibility of the utility to provide pure sinusoidal voltage. It suggests recommended practices to accomplish this.

3.3. Power Quality Improvement Techniques

The growing concern of customers and consideration of the economic factors give rise to the development of power quality improvement devices. Initially before the invention of power electronic devices saturable reactors, fixed capacitors and synchronous motors has been used as static and dynamic compensators respectively for improving the quality. Later FACTS devices of variable impedance and switched converter type replaced the conventional ones. In addition to FACTS devices the following methods like passive, active and hybrid filters are also in use for past two to three decades because of its merits compared with the other existing ones.

Passive Filters:

Passive filter comprised of passive elements like resistors, capacitors and inductors. Mainly they are categorized into two types: tuned and damped filters. Tuned filter can be series or parallel configurations. Series tuned filter offers low impedance at tuned frequency thereby bypassing the tuned harmonic frequency component through series resonance. Likewise, tuned filters can be connected to provide parallel resonance also. There it offers high impedance to the tuned frequency component. Auto tuned filter configuration is also available whose tuned frequency can be changed by switching necessary filter elements.

Damped filters are high pass filters. Unlike tuned filters it offers low impedance path for wide range of frequencies. Usually it has low Q factor values in the range of 0.5 to 5. Damped filters are classified into first order, second order, third order and C type filters.

Even though the filtering performance of third order is better than second order, for medium voltage application second order damped filters are used because of complexity of the circuit and reliability factors. The passive filters have the drawbacks like fixed compensation, bulky, dependency to power system network components, can offer resonance interacting with line components and certain cases it even acts as load. Also the filtering characteristics heavily dependent upon frequency variations, source impedance and tolerances in component values. Still it can be used in applications in which the cost and simplicity matters.

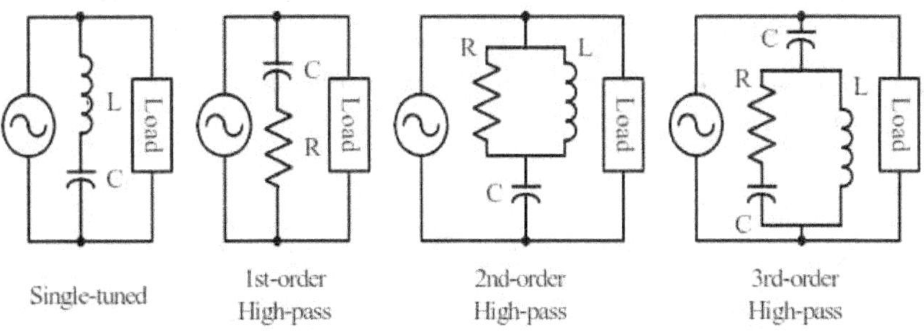

Fig 3.1 Common types of passive filters [8]

Active Filters:

The bulky structure, dependency to power system network parameters, incapabiliy to eliminate inter and sub harmonics, high losses etc has restricted the use of passive filters and which lead to the development of converter type compensators. Active Filters consist of power electronic devices which can be easily controlled to achieve required compensation. The performance charactristics of Active Filters are independent of power system network parametrs and independent control is possible.Active Filters can be voltage source inverter(VSI) type and current source inverter(CSI) type.VSI type filter is commomnly used considering the complexity,cost,reliability,protection and other implementation details.

Based on the topology ,Active Filters can be further categorized as Shunt and Series Active Filters.

Shunt Active Filters:

The Shunt Active Filters are commonly used for reactive power compensation, current harmonic eliminaion,balancing of load currents,flicker compensation,power factor improvement etc.If operate it in voltage controlled mode,it can even used for voltage regulation also.Fig.3.2 shows the VSI based Shunt Active Filter.The main components are VSI,DC link capacitance and coupling inductance.DC link capacitor provides input voltage to the inverter.The sensed parameters are given to the controller which generates the compensation currents with suitable control algorithms.The compensation currents can be injected to the line to provide necessary compensation.Coupling inductance helps to tune the comensation current.Controller generates the pulses to the inverter which drives the inverter switches.In current controlled mode Shunt Active

Filter acts as current source. Shunt Active Filter carries only compensation currents and loss component corresponds to meet the switching loss in inverter thereby reducing the rating of the filter.For high power rating several filters can be connected in parallel.

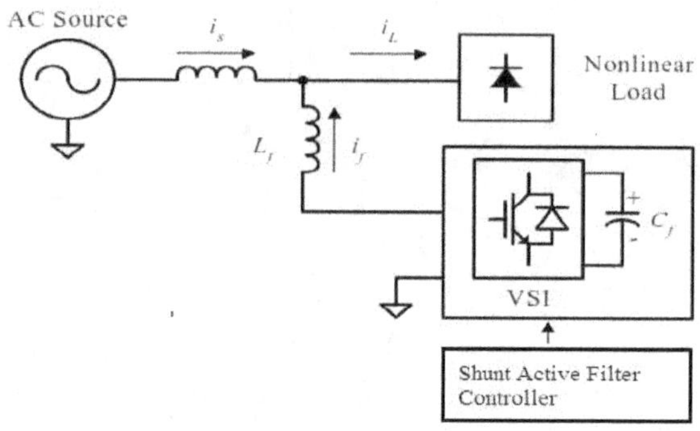

Fig 3.2 Shunt Active Filter [8]

Series Active Filters:

Series Active Filters are used to provide voltage balancing,voltage harmonic elimination ,elimination of voltage sag and swell etc.It is a crucial component to protect sensitive loads from voltage variations.These are connected in series with the line through injection transformers as shown in Fig.3.3.The VSI output will be the compensating voltage which will be added with the line voltage thereby maintains pure sinusoidal voltage with rated amplitude at the load terminals.Only draw back of Series Active Filter is that it has to carry the load current since it is connected in series with the line.Here the VSI operates in voltage controlled mode.

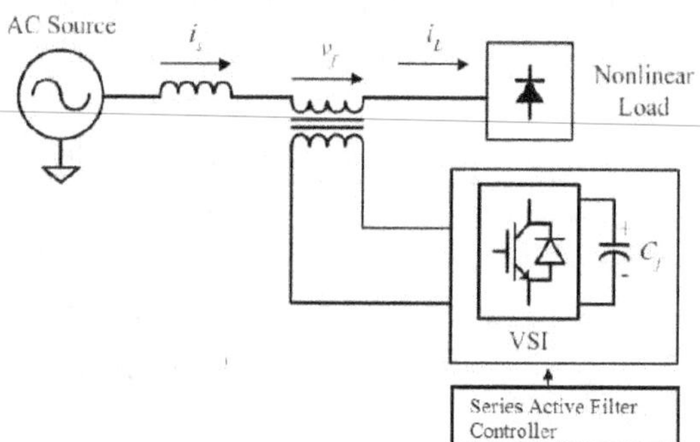

Fig 3.3 Series Active Filter [8]

To get the advantages of active and passive filters like improved filtering performance and cost reduction,combination of both active and passive filters can be used which is called as hybrid filters.There are various hybrid APFs have been proposed.Two commonly used configurations are shown in the Fig.3.4.UPQC is also a hybrid filter which is a combination of shunt active and Series Active Filters sharing a common DC link.All voltage and current related issues can be eliminated with the help of UPQC and also independent control of active and reactive power is possible.The configuration for UPQC is shown in Fig.3.5.

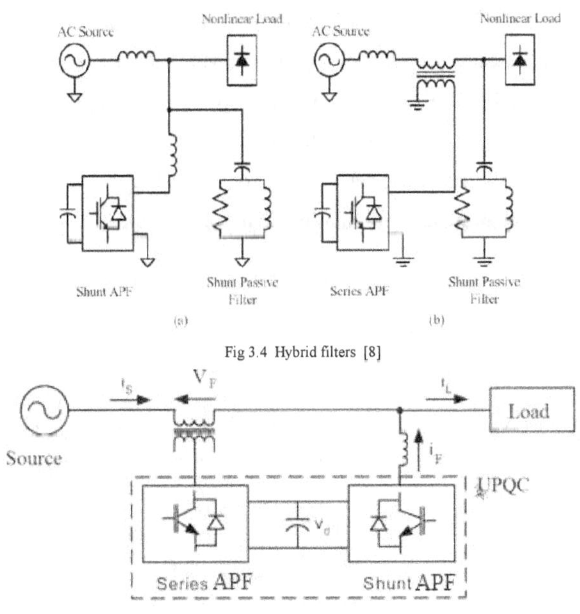

Fig 3.4 Hybrid filters [8]

Fig 3.5 UPQC

Another improvement technique is Improved Power Quality Converters(IPQC). The problems in conventional rectifier circuits can be eliminated using IPQCs. It can maintain constant voltage at dc side along with power factor improvement and current harmonic elimination. There are classifications based on converter type and topology based.

3.4. Scope for privatization of companies

In the Indian context, power generation and distribution sectors are largely controlled by government. The central government and state government owned electricity boards operate on power generation and distribution. Power distribution is the most important aspect in the whole process as it is the most vulnerable area considering the huge loss occurring to the distribution sectors. Even though various measures are taken by the government to reduce the losses, still power distribution sector remains the most vulnerable area. In such a context, privatization is the best solution. Privatization ensures increased productivity as the procedures are simpler than the usual government procedures and faster customer satisfaction as the whole survival depends on it. A Custom Power Park design is very much suitable for private investing groups as they will be having complete selection privileges for the fixing of rate and the entire market visibility is depending on the customer satisfaction. The choice of devices and control techniques decides the performance of the service provided by CPP. So this model suits small scale private investment teams to build CPP in large numbers across the country which will eventually boost the quality of power offered to the end users.

3.5. Conclusion

This chapter addresses the importance of power quality and its effects on the power systems. It also contains the improvement techniques implemented to maintain good power quality in power systems.

CHAPTER 4

CUSTOM POWER PARK

4.1. Introduction

In today's world, power system pollution has created a huge negative impact on economic aspect. Deregulated architecture with SCADA system itself is not able to address the consumer problems in full scale. The main reason behind India lagging behind in the case of power quality improvement is the common code of conduct between various DISCOMs Customers will be ready to pay little more if somebody else is ready to take the headaches like regular maintenance and servicing, installation, need of large space and technical support and high cost per unit of individual mitigation equipment. There comes the importance of a custom power park. Custom power parks buy grid power, improves the quality of power using various power conditioning equipment and sell it as different grades of power based on quality according to the requirement of the consumer. Here, consumers are getting an opportunity to select the kind of power they want. This chapter mainly focuses on the idea behind the custom power park and various custom power devices deployed in custom power park for its proper functioning.

4.2. Concept of Custom Power Park (CPP)

Custom Power Park (CPP) ensures the delivery of steady, clean, uninterrupted, high quality power to its customers according to their demand. The custom power park ensures the following characteristics:

- ➢ Voltage and Current distortions are within the limits specified by IEEE standard

- ➢ No surges or transients in the voltage

- ➢ Reliability

- ➢ Frequency and Voltage variations also will be within the limits specified by IEEE standard

- ➢ No Electro Magnetic Interference i.e, electromagnetic compatibility also according to IEC standards.

In CPP, the end users are getting an opportunity to select the kind of power according to their requirement. Power is fed to the park through two incoming feeders called as preferred and alternate feeders. To maintain the continuity of supply when the voltage is less than a certain limit, the switch over between preferred and alternate feeders are done using a solid-state transfer switch. In the distribution network, the power is divided into three different grades: Grade A, Grade AA and Grade AAA. The Grade A power itself is of improved quality without any current related issues like harmonics, low power factor and imbalance between phases, because of the service provided by a Shunt Active Filter. Grade AA power is slightly superior to Grade A category with the service provided by a diesel generator (DG). If both the feeder voltages are less than the limit or interrupted, then supply reliability can be maintained with the help of a DG. Grade AAA category will get services of the Shunt Active Filter, DG as well as a Series Active Filter. The time taken for the DG to come in synchronism with the grid is usually 10 to 20 seconds. During this period the service continuity for Grade A and Grade AA category will be temporarily lost, whereas it can be maintained for Grade AAA category with the help of Series Active Filter with battery support or renewable energy support even during that period. Thus, grade AAA category power will be of superior quality without any voltage or current related issues, thereby making it possible to maintain rated sinusoidal voltage at the load terminals under all conditions of supply. Grid integration of Renewable energy sources also can be done from the park through Shunt/Series Active Filter as interface to provide real power exchange. Fig.4.1 represents the block diagram representation of a Custom Power Park.

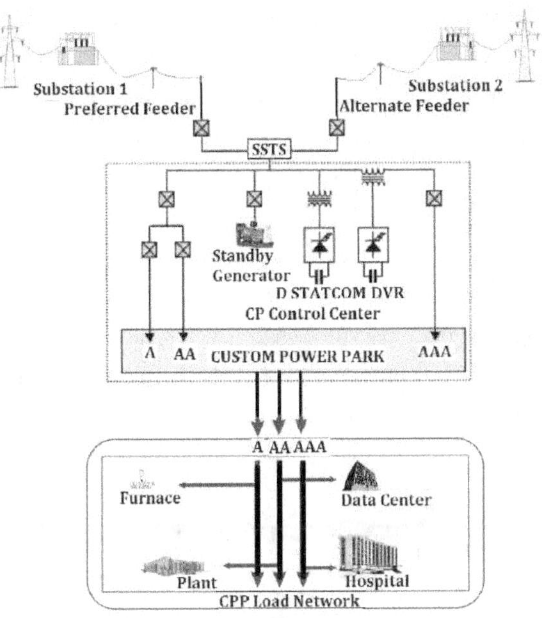

Fig.4.1 Block diagram representation of a Custom Power Park

4.3. Custom Power Devices

To ensure the delivery of high quality power to the consumers' certain power conditioning devices are installed in the custom power park so that the improvement in the quality of power supplied can be done. CPP concept apply not only to consumer loads but also to commercial and industrial loads connected to the distribution network.

The main power conditioning devices used in CPP are solid state transfer switch (SSTS), DSTATCOM/Shunt Active Filter, DVR/Series Active Filter, Diesel Generator (DG) and so on.

4.3.1. Solid State Transfer Switch (SSTS)

CPP Centre receives power from two different substations through preferred and alternate feeders. Usually the preferred feeder will be connected to the loads in the distribution network of CPP. The feeders are connected to the park through the SSTS, which is a pair of anti-parallel connected power electronic switches, so that the switching over between feeders can be initiated by appropriate control pulses. The SSTS transfers the load from preferred to alternate and vice versa in the occurrence of voltage variations due to any fault in the feeder. The commonly used switches can be made with thyristors, and GTOs depending on the transfer time, the control complexity and cost. With the application of negative gate current, GTO based switches will take approximately less than 5ms time to switch over compared to few milliseconds to half a cycle time taken by the thyristors. Owing to the fast transfer, GTO based switches ensure protection of different compensating devices in the event of a fault in the feeder by isolating them before the fault current reaches the limiting value. The main drawback of GTO based switches are the requirement of high magnitude negative current to turn off and also the cost. Thus, it is a tradeoff between fast transfer and cost-effectiveness. For GTO based switches since pulses are needed to be given to turn on and also to turn off, the complexity of the control circuit also needs to be accounted for.

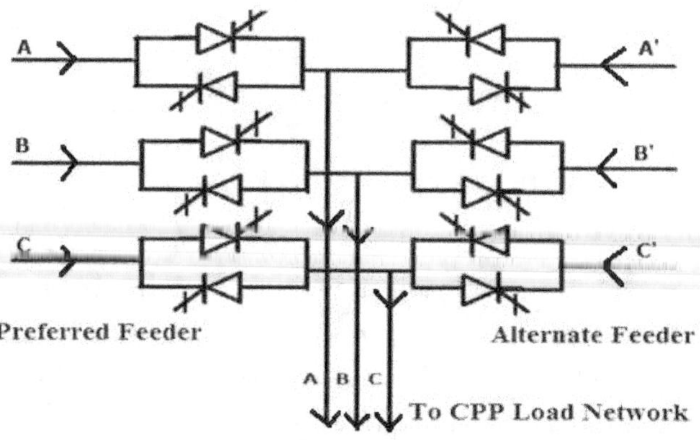

Fig.4.2.GTO based SSTS

4.3.2. DSTATCOM/Shunt Active Filter

Lately, the progress made in smart grids such as incorporation of various grid integrated renewable energy sources and development of charging stations of electric vehicles (EV), has also created power quality related issues in addition to the huge contribution from power electronic based loads. Renewable energy sources like solar, wind etc. can also influence the quality and reliability of the electrical network. By design, EV interface subsystems consist of power electronic converters that are non-linear devices with semiconductor-based switching elements. Thus, EV charging stations can create non-linear distortions and cause failure of electrical equipment.

To improve the transient stability in power grids, power flow needs to be controlled using power electronic switches. The device that is used for this purpose is categorized as Flexible AC Transmission System (FACTS) device and a static synchronous compensator (STATCOM) is a FACTS device. The basic functionality of a STATCOM is interconnection. This will secure the electrical grid and enables the continued adoption and use of renewable sources. It operates according to the voltage source Inverter (VSI) or current source inverter (CSI) principles and pulse width modulation, and has the ability to switch within milliseconds. Even if the system voltage drops, this is not affected by the magnitude of the voltage. This is a unique feature of STATCOM. It performs the functions like current harmonic elimination, reactive power compensation, power factor correction, balancing of load current etc. The block diagram representation of a Shunt Active Filter is shown in Fig.4.3.

The main parts of Shunt Active Filter are Voltage Source Inverter (VSI), Coupling reactance and/or shunt coupling transformer, control circuit and DC link capacitance.

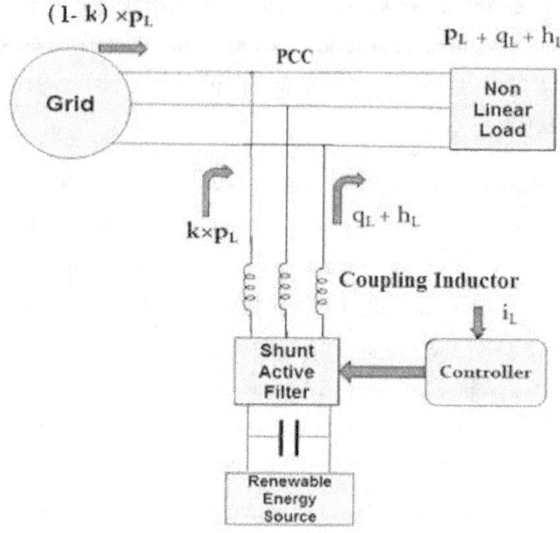

Fig.4.3 Block diagram representation of Shunt Active Filter

In Fig.4.3;

P_L-Real power demand from load

Q_L-Reactive power demand from load

h_L-harmonic component of load power

i_L-Load current

k-factor which represent the amount of power that need to be supplied by the source

PCC-Point of Common Coupling

The voltage source inverter acts as Shunt Active Filter whose output is controlled by pulses which are generated by comparing reference compensation currents with the actual compensation currents from the filter. Current source inverters also can be used as Shunt Active Filters. But due to control complexity and cost and many other factors, preference is given to voltage source inverters. Different configurations like three phase three wire, three phase four wire, multilevel etc are available. Commutating/coupling reactance is mainly used for current tuning so that the reference current tracks the actual one. Input to VSI is provided by dc link capacitor and the voltage change across dc link capacitance is an indication of real power exchange between the VSI and grid. The choice of the commutating reactance and dc link capacitance has significant effect on tracking the current and hence the compensation. High value of inductance makes tracking accurate but makes convergence slow. Similarly, high value of dc link capacitance reduces the ripple content in DC side more effectively while increases the cost. Low value of it increases the ripple content. There is always a tradeoff between the performance and cost. The control algorithm determines the necessary compensation current to carry out the functions like current harmonic elimination, reactive power compensation, and eradication of unbalance in the load currents by sensing parameters like voltage, current etc. If at the dc link a renewable energy source or a battery is connected, the Shunt Active Filter can also provide real power support to the load.

4.3.3. DVR/Series Active Filter

Voltage disturbances are a constant threat to the industries which deploy sensitive loads for their operations. When a voltage sag exceeds three cycles, it causes process interruption in many manufacturing industries. There can be an extension of process deadline due to reprogramming and restarting of sensitive equipment which are affected by the voltage variations. One of the devices used to protect these loads from voltage perturbations is the Dynamic Voltage Restorer (DVR)/Series Active Filter. It controls the voltage applied to the loads by injecting a voltage of compensating amplitude, phase and frequency in series with the distribution line. This operates as a filter between transmission line and end user thereby always providing clear, conditioned voltage to the loads. The performance of the DVR depends upon the rating of the load, type and duration of voltage disturbance and the magnitude and direction of the injected voltage.

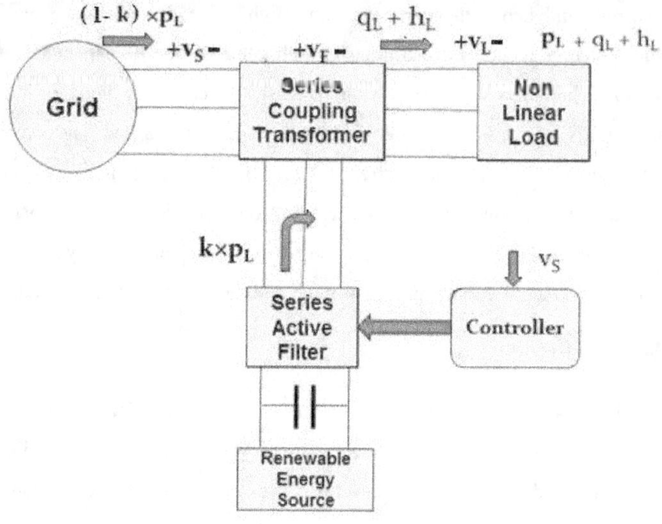

Fig.4.4 Block diagram representation of Series Active Filter

In Fig.4.4;

P_L-Real power demand from load

Q_L-Reactive power demand from load

h_L-harmonic component of load power

V_S-Source Voltage

V_L-Load Voltage

V_F-Filter Injected Voltage

k-factor which represent the amount of power that need to be supplied by the source

The main parts of Series Active Filter include Voltage Source Inverter (VSI), DC link capacitance, Coupling Inductor and/or series coupling transformer, control unit and harmonic LC filter. The role of coupling transformer is to inject three phase compensating voltages which contain fundamental as well as harmonic components in series with distribution line when a voltage disturbance occurs. The harmonic LC filter removes the ripples created in the output voltage of VSI due to high speed switching action of semiconductor devices. VSI employs fast active power electronic devices to convert DC to AC voltage and for that conversion the energy requirement is provided by DC link capacitor. The main function of control circuit is to continuously monitor the voltages and in case of any disturbance, generate reference filter voltages and pulses to inverter and also to maintain the DC link voltage at required constant level.

Series Active Filter injects voltage via series connected transformer which is added with line voltage to carry out necessary compensation. This injected voltage is generated by properly controlling the pulses which are generated by comparing the reference filter voltage generated by the controller and actual filter output voltages. The block diagram representation is shown in Fig.4.4. To ensure protection of sensitive loads from heavy sag/swell, a battery backup or a renewable energy source (RES) needs to be connected to the DC link. Real power exchange is also possible with this battery/RES support.

4.3.4. Diesel Generators (DG)/Renewable Energy Support

Custom Power Park is usually equipped with a Diesel Generator (DG) to maintain continuity of the supply to AA and AAA category loads when both preferred and alternate feeders fail. The DG takes about 10-20 secs to come to rated speed and synchronization. To eliminate this time delay, Shunt /Series Active Filters with RES support can be used to supply real power to loads for this short duration. For this purpose, renewable energy of any type like solar, wind etc can be used. According to the power rating of RES, it can transfer real power to AA and AAA category loads even for more duration. In that case, use of DG can be completely eliminated.

4.3.5. Hybrid Filter

Passive filters are usually employed to eliminate lower order harmonics by providing low impedance path at tuned harmonic frequency. It has advantages like less size, cost etc. Active filters have improved filtering characteristics compared to passive filters but are comparatively more expensive. Thus, to get the advantages of both active and passive filters, hybrid filter configurations are used. In that way the rating of the active filter can be reduced and the harmonic and reactive power compensation can be achieved in a cost-effective manner.

4.3.6. CPP Distribution Network

CPP basically delivers three grades of power such as Grade A, Grade AA and Grade AAA categories. This high-quality power is being achieved with the help of custom power devices installed in the park.

1) Grade A category power/A Loads: It gets the advantages of SSTS, which switches over from one feeder to other if fault occurs, or voltage goes below certain value. It also gets the services of the Shunt Active Filter thereby getting power with improved quality.

2) Grade AA category power/AA Loads: In addition to the services from SSTS and Shunt Active Filter, the continuity of the supply is ensured by DG connected in the park.

3) Grade AAA category power/AAA Loads: It receives the services of SSTS, Shunt Active Filter and Series Active Filter. Thus, even minute variations in voltage can be taken into account here thereby ensuring the protection of sensitive loads in the system. The load will get clean power without any sag, swell and harmonics along with reactive power support provided by Shunt Active Filter.

The examples of different categories of loads are shown in Table.4.1.

Table 4.1: Different categories of loads

Grade A	Office buildings, Computer hardware company, plastic company, consumer loads
Grade AA	Shopping malls, software development company
Grade AAA	Semiconductor chip manufacturing Industry, hospitals, data processing centers, biotech companies

4.4. Power Quality Improvement in Power Park using Custom Power Devices

The main custom power devices used for power quality improvement are the Shunt Active Filter and Series Active Filter. The role of these two devices in the CPP are explained below.

4.4.1. Power Quality Improvement using Shunt Active Filter/STATCOM

The main tasks performed by Shunt Active Filter are power factor correction, reactive power compensation, current harmonic elimination and balancing of unbalanced loads. Basically, Shunt Active Filters can be operated in two different modes:

(a) Current control mode: To achieve power factor correction, reactive power compensation, and harmonic elimination and load balancing, the filter has to be operated in current control mode. The load current contains fundamental and harmonic parts.

$$I_L = I_{fund} + I_h = I_{fp} + I_{fq} + I_h \tag{4.1}$$

where

i_L –Load Current

i_{fund}- Fundamental Component of Load Current

i_h- Harmonic Component of Load Current

i_{fp}- Active component (real part) of fundamental load current

i_{fq}- Reactive component (imaginary part) of fundamental load current

To achieve harmonic compensation, the filter has to provide the harmonic component requirement of the load. If reactive power compensation also needs to be performed along with harmonic compensation then the reactive component of fundamental also has to be supplied by the filter, so that at the source side unity power factor (UPF) can be maintained.

ie, compensating current, $I_F = I_{fq} + I_h$ $\tag{4.2}$

For linear loads the reactive component has to be provided by the filter in UPF mode of operation. The Shunt Active Filter will be able to feed the loads in CPP during the time taken by the DG for starting and synchronization if it is equipped with proper renewable energy support/battery backup.

$$P_L = P_{LDC} + P_{LAC} \tag{4.3}$$

where, P_L-Real power demand from load

P_{LDC}- DC component of real (active) power

P_{LAC} – AC component of real (active) power

The source will be required to provide only DC component of power which is nothing but the fundamental component of active power demand of the load. The AC component of active power and entire reactive power can be taken from Shunt Active Filter. With proper control techniques, even a fraction of fundamental active power can also be shared by the filter. Similar to reactive power compensation and harmonic elimination, the Shunt Active Filter can be deployed to eliminate unbalance in current too. For all these compensation techniques, the role of the controller is significant. There are time domain and frequency domain control algorithms reported in literature. A few of time domain control algorithms are as follows:

Synchronous reference frame theory (SRF), Instantaneous Reactive Power Theory (IRPT), Proportional integral controller based theory, Synchronous detection method, DC bus voltage algorithm, ICosΦ Control algorithm, Instantaneous Symmetrical Component Theory etc.

Fig.4.5 represents the operation of Shunt Active Filter under harmonic compensation.

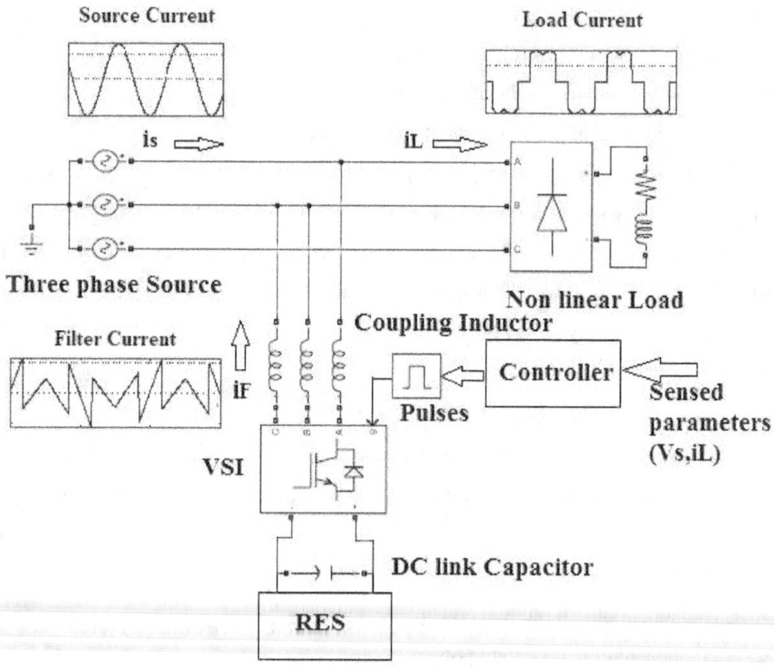

Fig.4.5 Block diagram representation of Shunt Active Filter under harmonic compensation

A diode bridge rectifier fed by a three-phase source is the system under consideration. The sensed inputs are given to the controller which determines the compensation currents. Switching pulses are then generated to drive the inverter switches and its output can be tuned by coupling inductor in current control mode of operation to achieve necessary compensation. The filter injects necessary compensation currents so that at source side the current is pure sinusoidal and at unity power factor even if the load is drawing distorted current which is clearly depicted in the diagram. *(b)Voltage Control mode*: In this mode the filter injects current in such a way that the terminal voltage follows a specified reference value. In steady state the filter should not inject or absorb any real power. The magnitude of injected current depends upon the rated voltage at load terminal whereas the angle is selected in such a way that the source needs to supply only real part of fundamental component of load current which is also strictly based upon the control algorithm. Whether in voltage control mode or current control mode, the performance of the Shunt Active Filter heavily depends upon the effectiveness of the control algorithms. In CPP, current control mode of operation is used to ensure reactive power and harmonic current compensation of all the categories of loads to improve the power quality.

4.4.2. Power Quality Improvement using Series Active Filter/DVR

In CPP, the Series Active Filter/DVR is used to ensure the protection of AAA category loads which are very sensitive to voltage variations. A minute voltage variation can cause large error, or loss of data or malfunctioning of equipment which together even contribute huge financial losses for the industries. Series Active Filter injects a dynamically controlled voltage in series with line to compensate the voltage perturbations. The main functions performed by Series Active Filter are voltage regulation, voltage sag and swell protection, voltage balancing and voltage harmonic elimination. All voltage related issues can be effectively compensated by the filter. The voltage at the load terminals, is continuously monitored and sensed and given to the controller which generates the necessary compensating voltage which is injected through the series connected coupling transformer.

$$V_l = V_s + V_f \tag{4.4}$$

Where V_l, V_s and V_f represents the load terminal voltage, source voltage and injected voltage respectively.

The Series Active Filter/DVR is considered as a voltage source of variable amplitude, frequency and phase angle. If the load terminal voltage is changed due to voltage disturbance, DVR injects necessary compensating voltage to restrain back the voltage to rated value. If the DVR needs to absorb/deliver only reactive power, then the positive sequence component of injected voltage has to be 90^0 out of phase with the positive sequence fundamental component of load current.

The Series Active Filter can be operated in standby mode and injection mode.

a) Standby mode: In standby mode, there is no need to inject the compensating voltage because already the load terminal is at rated voltage. Here, no pulses are given to the inverter which acts as the filter.

b) Injection mode: In this mode, the filter injects a three-phase voltage through coupling transformer due to the detection of voltage variation or harmonics. This injected voltage will be added with source voltage to provide rated, sinusoidal voltage at load terminals.

If DVR is equipped with RES support/Battery backup, real power exchange also can be done. This can have made use of in CPP to provide real power support to sensitive loads during the starting and synchronization time of the DG.

If there are only linear loads, then already four methods have been proposed for sag/swell compensation. They are pre-sag compensation method, in-phase compensation method, energy optimization compensation method and in-phase advanced compensation method. In pre-sag compensation method, the magnitude and phase angle, both are compensated. In the in-phase compensation method, the voltage is injected in phase with supply voltage. For certain loads the injected voltage magnitude will be minimum under this in-phase injection method. In energy optimization compensation method, the injected voltage will be in quadrature with supply voltage so that no active power exchange happens. In the in-phase advanced compensation method, the real power consumption is decreased by minimizing the angle between sagged voltage and load current.

However, in most of the cases, nonlinear loads will be there in the system and if percentage sag is more, the real power injection is also required. Under that case, the above-mentioned methods cannot be used. Various control algorithms have been proposed for the purpose. Some of them are IRPT and SRF based ones.

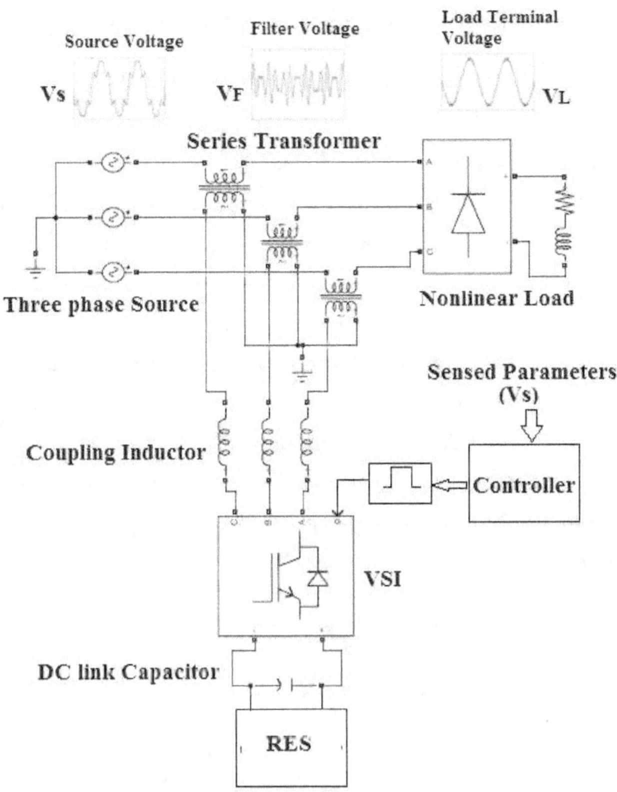

Fig.4.6. Block diagram representation of Series Active Filter with nonlinear load

Fig.4.6 represents the diagram which depicts the operation of Series Active Filter. The terminal voltage is continuously sensed and given to the controller. As shown in the diagram, because of the nonlinear currents drawn by the diode bridge rectifier load, harmonics are introduced in source voltage. Control algorithm calculates the necessary compensation voltages and injects them to the line by firing the switches of the inverter which when added with source voltage give rise to pure sinusoidal voltages of rated magnitude at load terminals, as depicted in Fig.4.6. Here also the performance characteristics depend upon the control algorithm.

4.4.3. Power Quality Improvement using Solid State Transfer Switch (SSTS)

SSTS ensures the protection of loads in CPP from sags/swells which arise due to faults in preferred and alternate feeders. If fault occurs in preferred feeder, then it will switchover from preferred to alternate feeder within 5-20 ms time with the help of efficient power electronic devices. The transfer control depends upon relative magnitude of the source voltages, feeder and load impedances, fault detection time, type of load etc. Proper design of transfer control is required to achieve less transfer time. Usually, make before break switching action is employed so that continuity of the supply will not be lost. But if there is fault in the feeder, this make before break switching strategy can lead to large currents to flow through the transfer switch. To overcome this difficulty, hybrid SSTS which is shown in Fig.4.7 can be employed. During normal operation, the mechanical switches will conduct and only when transfer is required, the thyristors will be switched on. Since the thyristors are coming into operation only during transient condition, the cooling requirement will be less and also the power losses can be minimized. For effective and fast control, the thyristors can be replaced with GTOs.

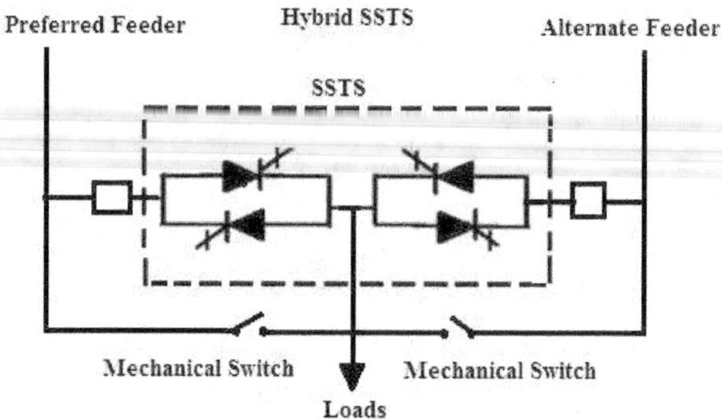

Fig.4.7. Thyristor based hybrid SSTS

4.5 Conclusion

The concept of custom power park and various custom power devices have been discussed in detail in this chapter. The power is mainly divided into three categories and any consumer can seek different categories of power. The role of different custom power devices for improving the quality of supply and to provide steady and clean power are vital.

CHAPTER 5

FUNDAMENTAL VOLTAGE PEAK DETECTION (FVPD) ALGORITHM FOR DVR/SERIES ACTIVE FILTER

5.1. Introduction

Control Algorithms play a major role in the performance of series active filters. This chapter presents the different control algorithms so far proposed for series active filters and compares the performance of the proposed controller with prevailing ones.

5.2. Different Control Algorithms for Series Active Filters

5.2.1. Instantaneous Reactive Power Theory (IRPT)

The instantaneous reactive power algorithm (IRPT) is an existing conventional algorithm [40-42] based on Clarke's transformation. The voltage and current vectors in phase coordinates can be transferred to stationary $\alpha\beta$ coordinates.

$$\begin{bmatrix} v_\alpha \\ v_\beta \end{bmatrix} = \sqrt{\frac{2}{3}} \begin{bmatrix} 1 & \frac{-1}{2} & \frac{-1}{2} \\ 0 & \frac{\sqrt{3}}{2} & -\frac{\sqrt{3}}{2} \end{bmatrix} \begin{bmatrix} v_{Sa} \\ v_{Sb} \\ v_{Sc} \end{bmatrix} \tag{5.1}$$

Where v_α and v_β represent the mains voltage at α and β axes and v_{Sa}, v_{Sb} and v_{Sc} represent the mains voltage in A, B and C phases respectively.

$$\begin{bmatrix} i_{L\alpha} \\ i_{L\beta} \end{bmatrix} = \sqrt{\frac{2}{3}} \begin{bmatrix} 1 & \frac{-1}{2} & \frac{-1}{2} \\ 0 & \frac{\sqrt{3}}{2} & -\frac{\sqrt{3}}{2} \end{bmatrix} \begin{bmatrix} i_{La} \\ i_{Lb} \\ i_{Lc} \end{bmatrix} \tag{5.2}$$

Where $i_{L\alpha}$ and $i_{L\beta}$ represent the load current in α and β axes and i_{La}, i_{Lb} and i_{Lc} represent the load current in A, B and C phases respectively.

The real and reactive power for the load can be calculated by the voltage and current in α β coordinates as:

$$\begin{bmatrix} p_L \\ q_L \end{bmatrix} = \begin{bmatrix} i_{L\alpha} & i_{L\beta} \\ i_{L\beta} & -i_{L\alpha} \end{bmatrix} \begin{bmatrix} v_\alpha \\ v_\beta \end{bmatrix} \tag{5.3}$$

p_L and q_L represent the instantaneous real and imaginary power consumed by the load. p_L and q_L contain both a DC(fundamental) term and an AC(harmonic) term. But for voltage harmonic suppression and deduction of voltage perturbations, the AC term of p_L and AC and DC terms of q_L should be provided by the Active Power Filter. Therefore, the reference signal of the compensation voltage can be represented as,

$$\begin{bmatrix} v_{f\alpha^*} \\ v_{f\beta^*} \end{bmatrix} = \begin{bmatrix} i_{L\alpha} & i_{L\beta} \\ i_{L\beta} & -i_{L\alpha} \end{bmatrix}^{-1} \begin{bmatrix} p_{LAC} \\ q_L \end{bmatrix}$$

(5.4)

The reference signal of the compensation voltage is obtained by transforming $V_{f\alpha^*}$ and $V_{f\beta^*}$ to three phase system using inverse Clarke's transformation.

$$\begin{bmatrix} v_{fa^*} \\ v_{fb^*} \\ v_{fc^*} \end{bmatrix} = \sqrt{\frac{2}{3}} \begin{bmatrix} 1 & 0 \\ \frac{-1}{2} & \frac{\sqrt{3}}{2} \\ \frac{-1}{2} & \frac{-\sqrt{3}}{2} \end{bmatrix} \begin{bmatrix} v_{f\alpha^*} \\ v_{f\beta^*} \end{bmatrix}$$

(5.5)

Where V_{fa^*}, V_{fb^*} and V_{fc^*} represent the reference filter voltages in a, b and c phases respectively.

The block diagram representation of IRPT controller is shown in Fig.5.1.

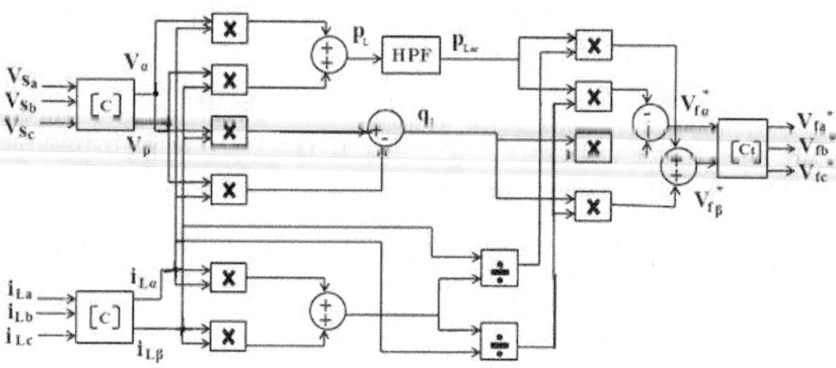

Fig.5.1. Block diagram representation of IRPT controller

The sensed load currents and voltages are given as inputs to the controller. Using the mathematical equations given in equation 5.1 and 5.2, the corresponding voltages and currents in stationary reference frame can be obtained. From these values, the active and reactive powers are calculated and the AC part of active power is extracted using a high pass filter. The compensation voltages are then calculated using the relation as shown in equation 5.4, which is later transformed back to three phase system using inverse Clarke's transformation.

5.2.2. Synchronous Reference Frame (SRF) Theory

In Synchronous-Reference-Frame-based control algorithm, initially sensed voltages are transformed to dq axis using Park's transformation

$$\begin{bmatrix} v_d \\ v_q \\ v_0 \end{bmatrix} = \frac{2}{3} \begin{bmatrix} cos\theta & -sin\theta & \frac{1}{2} \\ cos(\theta - 120^0) & sin(\theta - 120^0)\frac{1}{2} \\ cos(\theta + 120^0) & sin(\theta + 120^0)\frac{1}{2} \end{bmatrix} \begin{bmatrix} v_{sa} \\ v_{sb} \\ v_{sc} \end{bmatrix} \tag{5.6}$$

where, v_d and v_q represent supply voltages in d and q axes with DC and AC parts as given below, and v_a, v_b and v_c represent supply voltages in A, B and C phases respectively.

$$V_d = V_{dDC} + V_{dAC} \tag{5.7}$$

$$V_q = V_{qDC} + V_{qAC} \tag{5.8}$$

The AC part of v_d and v_q are extracted using a low pass filter. The load voltage magnitude is subtracted from the desired amplitude and the error in load voltage magnitude is fed to a PI controller. The PI controller output corresponds to the reactive component of voltage, for maintaining constant terminal voltage. PI controller output is then added with V_{qDC}.

$$V_{q*} = V_{qDC} + V_{qr} \tag{5.9}$$

where V_{qr} is the PI controller output.

V_d and V_{q*} are converted back to three-phase to get reference source voltages(Vs^*). Reference filter voltages(V_f^*) can be calculated by subtracting sensed load voltages(V_L) from reference source voltages.

Block diagram representation of SRF control algorithm is given in Fig.5.2. Here, self-supported DC bus is not considered. To make a self-supported DC bus, the corresponding loss component as shown in equation (5.10) can be generated using PI controller, which needs to be added with V_{dDC}

$$V_{loss(n)} = V_{loss(n-1)} + K_p(V_{DCerror(n)} - V_{DCerror(n-1)}) + K_i V_{DCerror(n)} \qquad (5.10)$$

Where

$V_{DCerror(n)} = V_{DC}^* - V_{DC(n)}$, is the error between the reference DC bus voltage and sensed DC bus voltage at nth sampling instant.

$V_{loss(n)}$ and $V_{loss(n-1)}$ - loss component at n^{th} and $(n-1)^{th}$ sampling instant

K_p, K_i - proportional and integral gain of PI controller

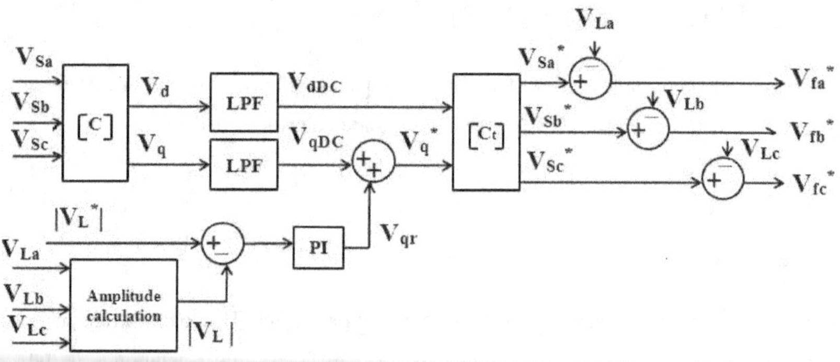

Fig.5.2. Block diagram representation of SRF controller

Sensed voltages (Vsa, Vsb, Vsc) are converted to rotating reference frame by Park's transformation. The reactive component of voltage (Vqr) required to maintain the load terminal voltage at rated value can be obtained by passing the error between sensed load voltage and rated load voltage magnitude   through PI controller. The sum of fundamental component (V_{dDC} and V_{qDC}) and the reactive component of voltage for voltage control across the load terminal (Vqr) represents the reference voltages at PCC (Vsa*, Vsb*,Vsc*) which is obtained by doing inverse Clarke's transformation of Vq*and V_{dDC}. The reference filter voltages(V_f*) can be calculated by subtracting sensed load voltages from reference voltages at PCC.

5.2.3. Proposed Fundamental Voltage Peak Detection (FVPD) Control algorithm

The IRPT and SRF based control algorithms include complex calculations which lead to computational overheads. To overcome such disputes and to improve the response time, a cost-effective solution is developed known as the FVPD controller. The main stages in the control algorithm starts with extracting the fundamental part of detected source voltage and measurement of its peak value. Hence the name Fundamental Voltage Peak Detection (FVPD) algorithm.

proposed control algorithm generates reference voltage signals, by sensing voltage supplied by the utility to load. This reference voltage is compared with actual voltage to generate pulses for the inverter, which acts as the Series Active Filter.

Distorted voltages in a, b and c phases supplied by the utility to the load can be represented as follows:

$$V_{Sa} = \sum_{n=1}^{\infty} V_{San} Sin.\,n\omega t \tag{5.11}$$

$$V_{Sb} = \sum_{n=1}^{\infty} V_{Sbn} Sin.\,n(\omega t - 120^0) \tag{5.12}$$

$$V_{Sc} = \sum_{n=1}^{\infty} V_{Scn} Sin.\,n(\omega t - 240^0) \tag{5.13}$$

Source voltage contains both fundamental as well as harmonic components. Extracting the fundamental components using low pass filter with cut off frequency of 50Hz,

$$V_{SaF} = V_{Sa1} Sin.\,\omega t \tag{5.14}$$

$$V_{SbF} = V_{Sb1} Sin.\,(\omega t - 120^0) \tag{5.15}$$

$$V_{ScF} = V_{Sc1} Sin.\,(\omega t - 240^0) \tag{5.16}$$

Dividing fundamental components by peak value, unit amplitude sine wave can be obtained.

$$U_a = \frac{V_{SaF}}{V_{Sa1}} sin.\,\omega t \tag{5.17}$$

$$U_b = \frac{V_{SbF}}{V_{Sb1}} sin.\,(\omega t - 120^0) \tag{5.18}$$

$$U_c = \frac{V_{ScF}}{V_{Sc1}} sin.(\omega t - 240^0) \tag{5.19}$$

Where U_a, U_b, U_c represents unit amplitude sine waves in respective phases.

If voltages in the three-phases do not have exact 120^0 phase shift between them, then U_a can be taken as reference, and U_b and U_c can be generated, from U_a, by phase shifting it with 120^0 and 240^0 respectively.

$$U_b = U_a \angle - 120^0 = sin(\omega t - 120^0) \tag{5.20}$$

$$U_c = U_a \angle - 240^0 = sin(\omega t - 240^0) \tag{5.21}$$

The reference source voltages can be generated, by multiplying unit amplitude sine waves with rated value of the voltage. Let V be the rated peak value then,

$$V_{Saref} = V \times U_a = V.Sin\omega t \qquad (5.22)$$

$$V_{Sbref} = V \times U_b = V.Sin(\omega t - 120^0) \qquad (5.23)$$

$$V_{Scref} = V \times U_c = V.Sin(\omega t - 240^0) \qquad (5.24)$$

Where V_{saref}, V_{sbref}, and V_{scref} represents reference source voltages which is nothing but the ideal voltage at PCC. Reference filter voltages can be deduced by subtracting the sensed source voltages from reference source voltages and represented as,

$$V_{Fa^*} = V_{Saref} - V_{Sa} \qquad (5.25)$$

$$V_{Fb^*} = V_{Sbref} - V_{Sb} \qquad (5.26)$$

$$V_{Fc^*} = V_{Scref} - V_{Sc} \qquad (5.27)$$

Block diagram representation of the proposed controller is given in Fig.4.

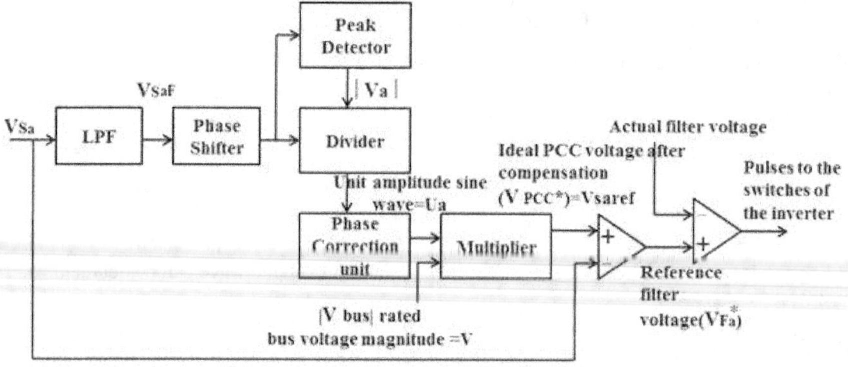

Fig.5.3. Block diagram representation of proposed FVPD controller for one phase

Sensed Source voltage is passed onto the second order bi-quad filter, to extract the fundamental component. The bi-quad filter output has an inherent 90^0 phase shift, which can be removed, using a phase shifter. Unit amplitude sine wave is obtained by dividing phase shifter output by the peak value, generated by the peak detector circuit. Phase correction unit provides the necessary phase shift for B and C phases, by taking 'A' phase as reference. The unit amplitude sine waves of the three phases are multiplied with rated magnitude to get reference source voltages which represents the ideal PCC voltage after compensation. The sensed source voltages are subtracted from the reference source voltages to obtain reference filter voltages. These reference filter voltages, and sensed actual filter voltages, are fed to a comparator to generate the pulses, for the inverter, which is used as series active filter to carry out the compensation.

5.3. System with proposed FVPD controller for simulation

Fig.5.4 shows a schematic diagram of a three-phase system comprised of a three-phase voltage source with a Series Active Filter, feeding a diode bridge rectifier/non-linear load. The nonlinear current drawn by diode bridge rectifier load, causes distortions of voltage at the load terminals, which can be rectified by switching on the Series Active Filter. A voltage source inverter (VSI) is used as the Series Active Filter, which injects the compensation voltage, through a coupling transformer in series with the line, to achieve the required compensation. In a Series Active Filter, it is the controller that generates accurate reference filter voltage templates for the Series Active Filter, to switch accordingly, and generate the necessary compensation action.

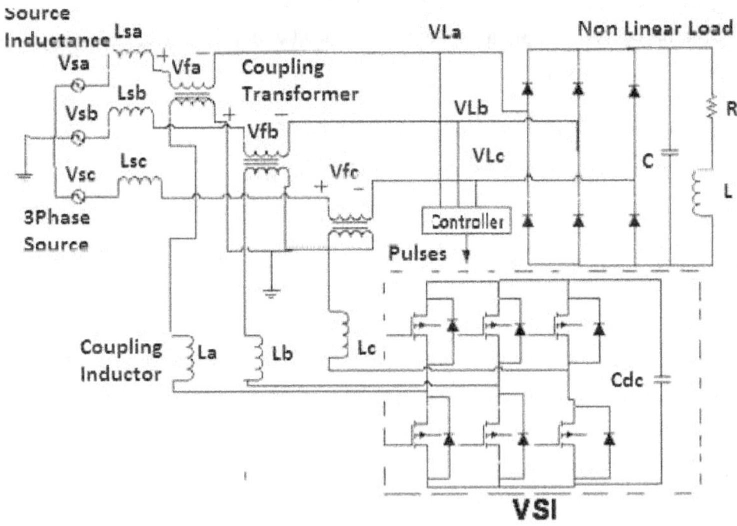

Fig.5.4. Series Active Filter connected to three phase system

The Design considerations for the simulation analysis is shown below.

5.4 Design Considerations

a) kVA rating of VSI

For harmonic compensation, the injected voltage $V_{inj} = \sqrt{V_5^2 + V_7^2 + V_{11}^2 + V_{13}^2 + ..}$

In the simulations, fifth and seventh harmonics of 20 % and 15 % of magnitude of the rated voltage were considered. All the other harmonics were neglected due to smaller magnitude compared with the fifth and seventh.

For harmonic compensation, the injected voltage $V_{inj} = \sqrt{V_5^2 + V_7^2}$

Considering the worst case of sag and swell and above mentioned fifth and seventh harmonics together,

total injected voltage for harmonic and sag swell compensation $= \sqrt{V_{sag}^2 + V_5^2 + V_7^2} = 207$ V

where Vsag is the voltage to be injected under sag/swell case alone.

Current rating = load connected downstream = 7.46 A for a load of 5 kVA

kVA rating = $(3 \times 207 \times 7.46)/1000 = 4.4$ kVA

b) kVA rating of Injection transformer

kVA rating of Coupling/Injection Transformer= $(3 \times 207 \times 7.46)/1000 = 4.4$ kVA

c)DC Capacitor voltage of VSI

$V_{DC} > 2\sqrt{2} V sc$ where Vsc is the secondary side voltage

d)DC bus capacitance of VSI

Considering that the energy stored in the capacitor is for meeting the energy demand of the load, for a fraction of a cycle, the relation can be written as

$$0.5\, C\, (V_{DC}^2 - V_{DCdelta}^2) = 3 V_c I_s\, \Delta t \tag{5.28}$$

Where V_c and I_s are the voltage rating of the filter and current through it, $V_{DCdelta}$ is the change in the DC bus voltage, during transient condition, and Δt is the time for which support is required

e) Design of Interfacing Inductor

Based on the ripple current in Series Active Power Filter, the value of the inductor can be determined

$$L = n\left(\frac{\sqrt{3}}{2}\right) mV_{dc}/(6af_s\Delta I_s) \qquad (5.29)$$

Where, n = turns ratio=1

ΔI_s= ripple current in the inductor taken as 5%

m = modulation index=0.8

a = overloading factor=1.2

f_s = switching frequency= 20kHz

The design values are given in Table 5.1.

Table 5.1: Design Values

Load	5kW,100VAr
kVA rating of Inverter	4.5 kVA
kVA rating of Transformer	4.5 kVA
V_{dc}	650V
DC link capacitor	9mF
Interfacing Inductor	25mH

5.5 Simulation Analysis of proposed FVPD control Algorithm

The MATLAB/Simulink circuit for the proposed system is shown in Fig.5.5. A three-phase source with 400V, 50Hz is considered as the supply system which delivers power to a three phase RL load of 5kVA(5kW,100VAr) rating. Voltage disturbances are introduced in the voltage using a three-phase programmable voltage source to analyze the performance of the controller during these abnormalities. Three-phase voltage source inverter is used as active filter which is connected to the line through series connected coupling transformers. The source voltage is sensed and given to the controller to generate compensating voltages. By using hysteresis controller, pulses to the inverter are generated by comparing actual and reference filter voltages which in turn drives the inverter switches to produce the voltage waveform apt to carry out necessary compensation. Simulation is also done by connecting diode bridge rectifier as nonlinear load.

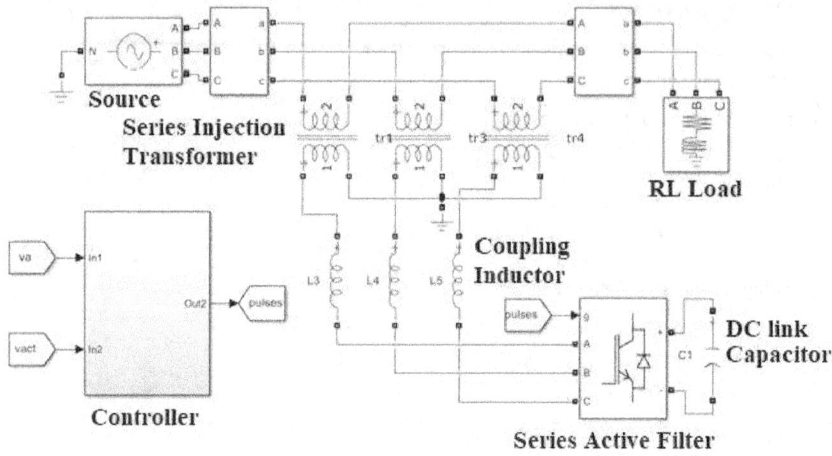

Fig.5.5. Simulation Circuit of series active filter connected to the system

To analyze the performance of the controller different cases like harmonics, sag, swell, magnitude unbalance and phase unbalance are considered. The simulation circuit for the controller is shown in Fig.5.6.

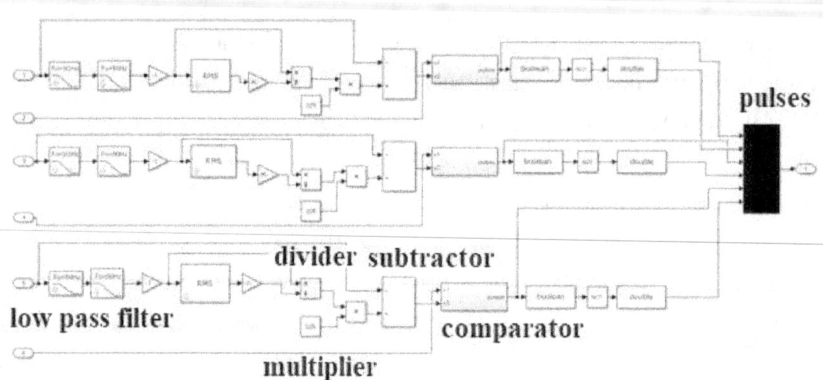

Fig.5.6. Matlab Simulink Circuit of Controller

a) Controller performance during Harmonics, Sag and Swell:

Fifth and seventh harmonics are introduced using programmable voltage source during the period 0.05s to 0.1s. Similarly, a three phase voltage swell of 20% is introduced from 0.1s to 0.15seconds. A three phase voltage sag of 20% has created from 0.15s to 0.2s. Fig 5.7 represents the controller waveforms under harmonics, sag and swell conditions respectively. The first component of controller is a low pass filter. The sensed source voltage which is the first waveform in Fig.5.7, is given to low pass filter to extract the fundamental component. The output of biquad filter is the 50Hz component with same magnitude as that of sensed voltage with an inherent phase shift of 90^0 as shown as second waveform in the Fig.5.7. The phase shifter removes the extra 90^0 shift and peak detector determines the peak value as represented by third and fourth waveforms. Fifth waveform represents the output of the divider which is nothing but unit amplitude sine wave which is then given to phase correction unit which maintains 120^0 phase shift between the phases in the case of any unbalance in the phase and is represented by sixth one. The output of phase correction unit is then multiplied by rated magnitude to get the reference source voltage as represented by the seventh waveform in Fig.5.7, which is nothing but the expected voltage at PCC. Reference filter voltage can be obtained by subtracting sensed voltage from reference source voltage represented by the last waveform.

It is also observed that the reference filter voltage in the case of sag is in phase with system voltage so that once it is added, the voltage at the load terminal can be boosted up to rated value whereas in the case of swell, it is just the opposite, i.e. 180^0 out of phase with the system voltage.

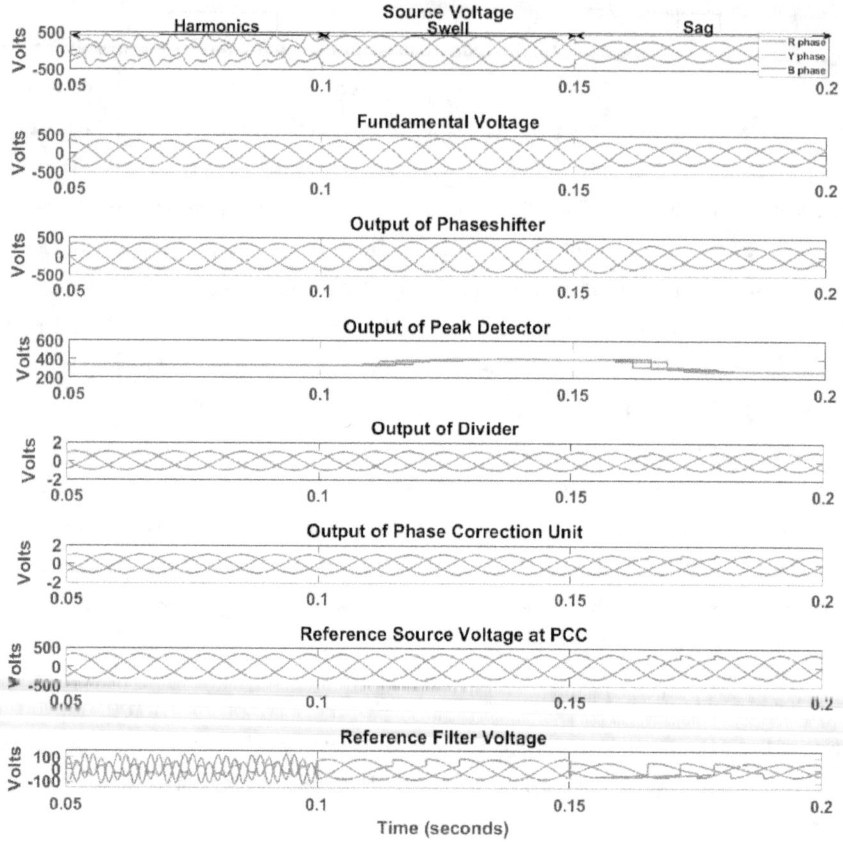

Fig.5.7. Controller waveforms during harmonics, Sag and Swell

b) Controller performance during Magnitude and Phase Unbalance:

To create a magnitude unbalance, the voltage at phase A is increased to 20% of rated within the duration 0.25s to 0.3s and decreased to 20% for a period 0.3s to 0.35s. The phase unbalance is created by introducing 20^0 extra phase-shift between the phases, i.e. the phase shift between A and B phases is 140^0 and between A and C phases is 260^0 from 0.35s to 0.4s. Fig.5.8 represents the controller waveforms during magnitude and phase unbalance.

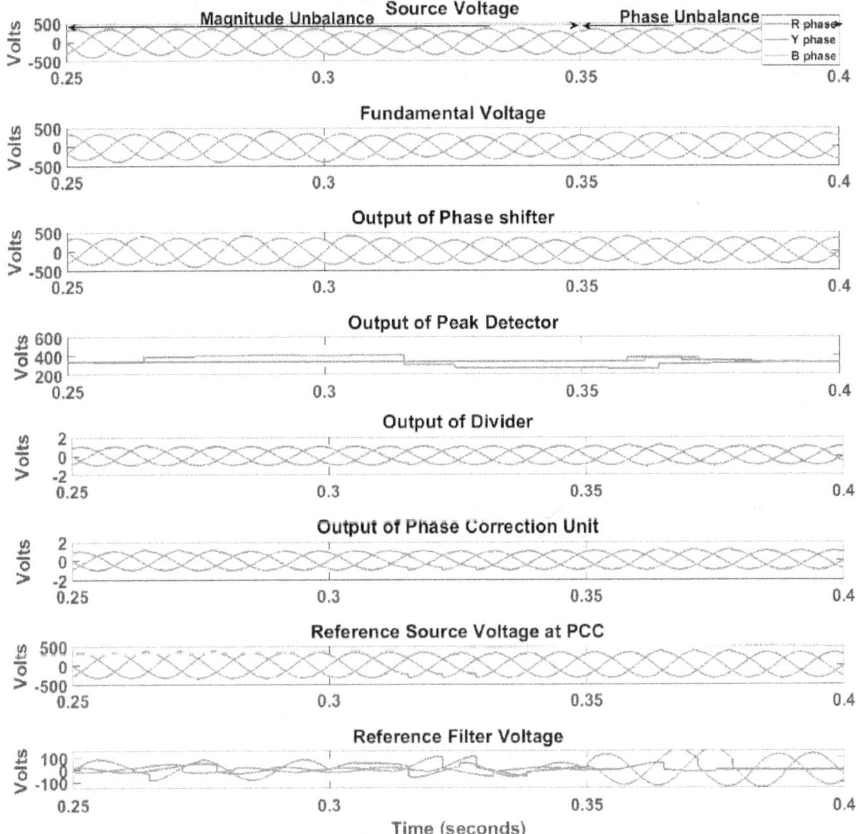

Fig.5.8. Controller waveforms during magnitude and phase unbalance

It is seen from Fig.5.8 that during magnitude unbalance the reference voltage is generated in phase A only since the other phase voltages are at rated value. Also, it is clearly seen that the phase correction unit eliminates the extra 20^0 phase shift and maintains 120^0 phase shift between the phases. Since A phase is taken as reference for the phase correction unit, the reference voltages are generated for B and C phases so as to do necessary compensation.

Fig.5.9 represents the source voltage and voltage at load terminals after the compensation for harmonics, voltage swell, voltage sag and magnitude and phase unbalance conditions respectively.

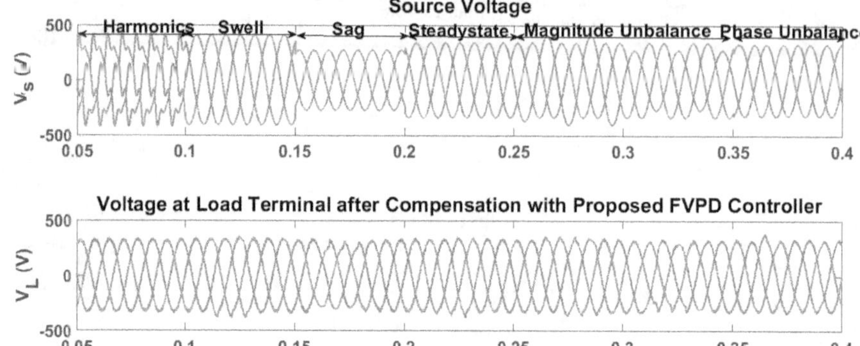

Fig.5.9. Source Voltage and Voltage at Load terminal after Compensation

The durations of each cases are given in Table 5.2.

Table 5.2: Voltage conditions and duration

Voltage condition	Duration in seconds
Harmonics	0.05-0.1
Swell	0.1-0.15
Sag	0.15-0.2
Steady state	0.2-0.25
Magnitude Unbalance	0.25-0.35
Phase Unbalance	0.35-0.4

Under all these voltage perturbations the voltage at the load terminals is maintained as a pure sinusoid with rated amplitude with the action of proposed FVPD controller. This clearly depicts the effectiveness of the controller to eradicate all the voltage abnormalities thereby ensuring the protection of sensitive loads from all voltage related issues.

In the previous circuit (Fig.5.5) a programmable voltage source is used to create harmonics, sag and swell. To consider the real time application, simulation circuit is made with diode bridge rectifier (DBR) as the load, as seen in Fig.5.10. Initially the 11kV voltage is boosted to 132kV by step up transformer and is transmitted via pi model transmission line. At the receiving end, the voltage is again stepped down to 11 kV with the help of two step down transformers and one feeder is connected to RLC load. Another11kV feeder is connected to DBR as well as RLC load. In this feeder a three phase fault creates sag and the nonlinear current drawn by the DBR distorts the voltage.

The associated waveforms are shown in Fig.5.11.

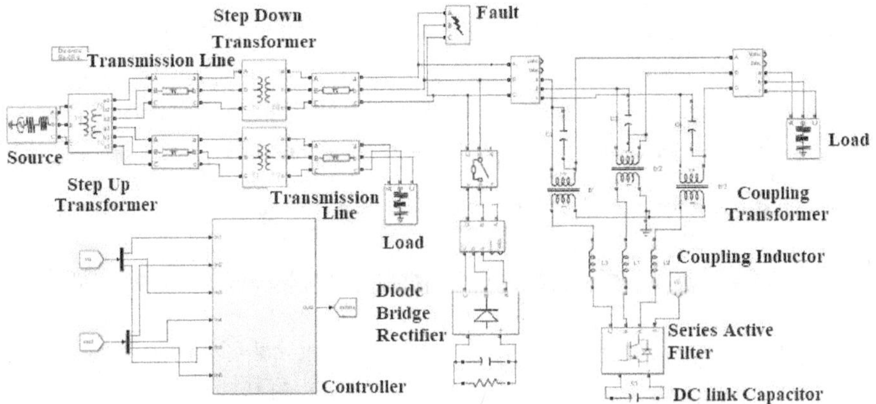

Fig.5.10.Simulation Circuit with diode bridge rectifier (DBR) as the load

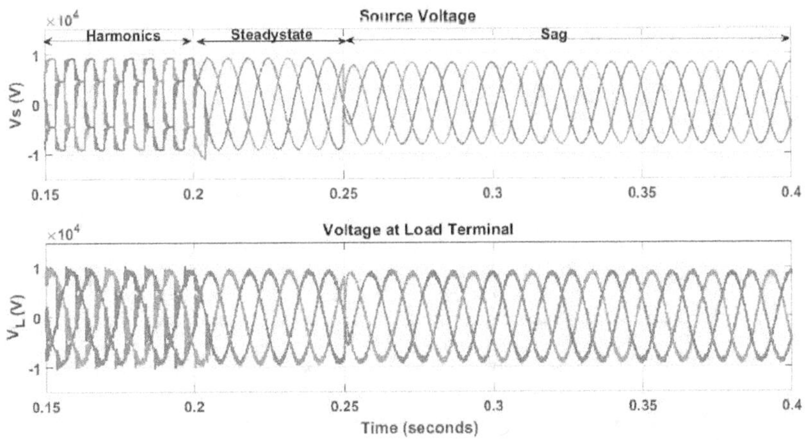

Fig.5.11. Source voltage and voltage at load terminals with FVPD controller

The DBR is switched at 0.15 sec till 0.2 sec and fault is introduced from t=0.25sec to t=0.4 sec. From Fig.5.11.it is seen that the distortion in voltage due to nonlinear current drawn by the DBR is reduced due to action of proposed controller. Also the sag caused by the three phase fault is also compensated by the filter with FVPD controller.

In Fig.5.11, it is observed that full compensation in voltage profile is not happening. This is due to the fact that the diode bridge rectifier is drawing a highly non-linear current. However, in the actual CPP implementation which is explained in section 6.4.2.1, good voltage compensation is taking place even when the diode bridge rectifier is drawing non-linear current, as current compensation is carried out by Shunt Active Filter additionally. Thus, pure sinusoidal voltage is maintained at AAA load terminal in CPP network even in the presence of diode bridge rectifier.

5.5.1. Simulation of Series Active Filter with fuzzy based controller

The backbone of FL control technique is the fact that it recreates the human potential to figure out the system's behavior. Based on this understanding, quality control rules are constructed. The important concept of FL is the fact that it arrives to a conclusion which is definite in nature from blurred, ambiguous, imprecise, noisy, or missing input data. [63-67] The basic structure of an FLC is represented in Fig.5.12.

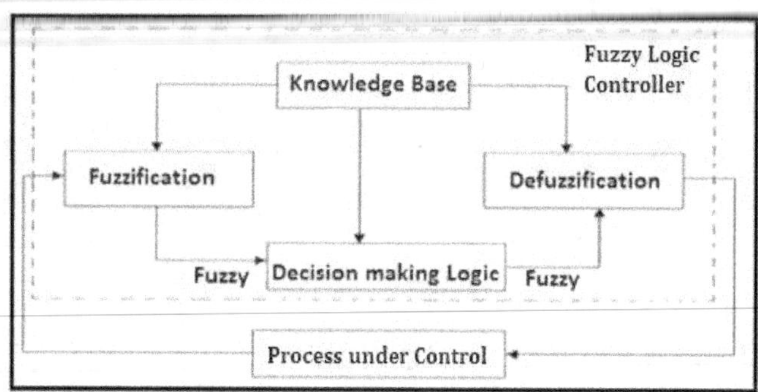

Fig.5.12. Basic block diagram of Fuzzy logic controller

5.5.1.1. Constructing a Fuzzy Controller

The main parts involve fuzzifier, knowledge base and rule base, decision making unit and defuzzifier. Since the fuzzy inference system process only fuzzy sets, the set of crisp inputs need to be converted to fuzzy sets with appropriate membership functions. This function is performed by the fuzzifier. The knowledge base is basically a database of fuzzified inputs (linguistic variables). Rule base contains set of defined rules. Based on the datas in knowledge and rule base, the decision making is done by the fuzzy inference system based on mamdani or takagi sugeno models and produce the output. This fuzzified output is converted to crisp output with the help of defuzzifier for necessary control action.

The calculated error from the measured and reference voltage is the first input to the fuzzy controller. The other input is change in error. The output is the reference filter voltage. The range for error input is -60 to 60V and range for change in error is -20 to 20V.Seven membership functions (NL, NM, NS, Z, PS, PM, PL) have been considered for both the inputs. For output, the range was -120 to 120V with nine membership functions. The membership functions for inputs and output, and the rule surface is shown in Fig.5.13.

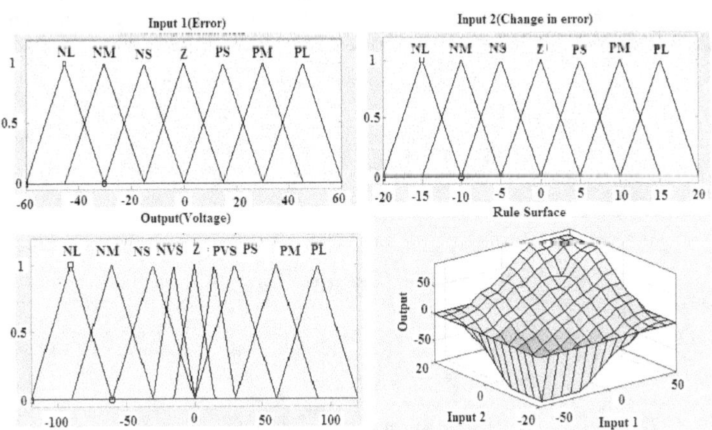

Fig.5. 13. Membership functions are rule surface

Initially the system was simulated with proposed FVPD controller. The rules are made with respect to the data obtained with the proposed FVPD controller. In total there are 49 rules which are shown in Table 5.3.

Table 5.3: Rule Table

e \ Δe	NL	NM	NS	Z	PS	PM	PL
NL	PL	PL	PL	PM	PS	PVS	Z
NM	PL	PL	PM	PS	PVS	Z	NVS
NS	PL	PM	PS	PVS	Z	NVS	NS
Z	PM	PS	PVS	Z	NVS	NS	NM
PS	PS	PVS	Z	NVS	NS	NM	NL
PM	PVS	Z	NVS	NS	NM	NL	NL
PL	Z	NVS	NS	NM	NL	NL	NL

The simulation results are shown in Fig.5.14. Performance of the controller under different percentages of sag and swell (0 to 25%) has been considered for simulation analysis and the waveforms are shown in Fig.5.14. When the voltage is in steady state (rated voltage during 0.15 to 0.2sec) the compensating voltage is zero. It is observed that whenever there is voltage variation, the filter injects compensating voltage thereby maintaining load terminal voltage at its rated value.

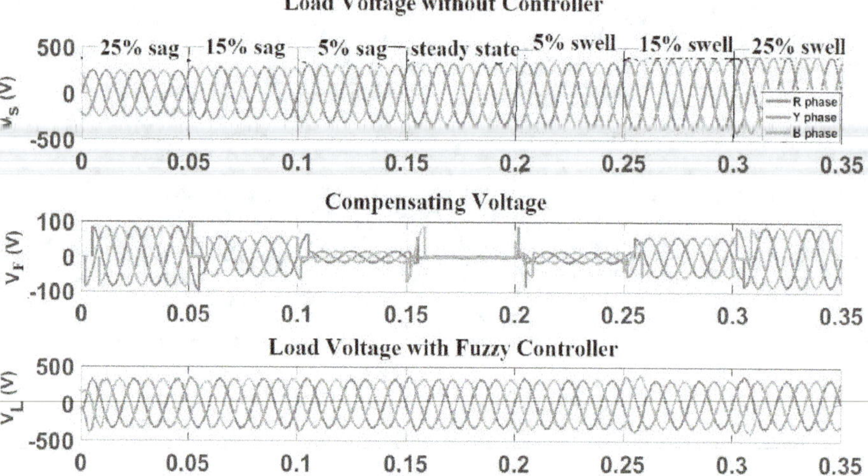

Fig.5.14. Load Voltage without fuzzy based controller, Compensating Voltage and Load Voltage with fuzzy controller for different values of sag and swell

The performance of the controller under magnitude unbalance condition is also tested and results are shown in Fig.5.15. From 0 to 0.05sec, 20% sag is created in 'phase A' and from 0.1 to 0.15 sec 20% swell is introduced in the same phase whereas the voltages in the other two phases remain at rated value. Under both these conditions of unbalance, the three phase voltages at load terminal is maintained constant with proper control action performed by the fuzzy based controller. During magnitude unbalance condition the compensating voltage in the other two phases are zero.

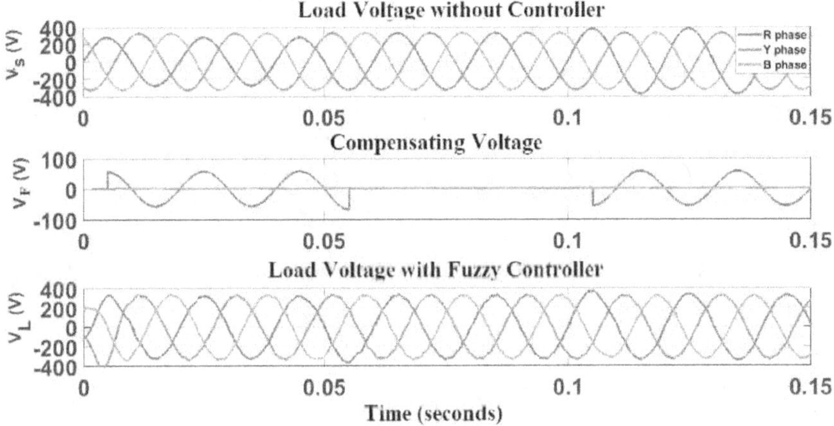

Fig.5.15. Load Voltage without fuzzy based controller, Compensating Voltage and Load Voltage with fuzzy controller under magnitude unbalance

5.6 Hardware implementation of proposed FVPD control Algorithm

5.6.1. Digital Controller

Even though analog circuits demonstrate its high accuracy and instantaneous response, over a period of continuous usage, malfunctioning of these devices are often noticed. These malfunctions are mainly due to wearing out of the components caused by heating and offsets in electronic circuits. But digital circuits and digital processors have demonstrated capabilities such as ease of use, durability and reliability. Due to these advantages, the proposed control algorithm is implemented digitally, by replacing all the analog circuits by digital controller. This digital implementation is done using TMS320F28335 microcontroller. To program this controller, Code Composer Studio v4 (Ccsv4) is used [80]. .

TMS320F28335

TMS320F28335 DSP Processor which comes under the C2000 series of 32-bit micro controller devices manufactured by Texas Instruments is picked to develop the controller. C2000 devices are 32-bit microcontrollers which exhibit high performance with the help of integrated peripherals designed for real-time control applications. Multiple complex algorithms can use the core present in the microcontroller at speeds necessary for demanding control applications. Powerful integrated peripherals combined with the SPI, UART (SCI), I2C, CAN, and McBSP communication peripherals make C2000 devices the perfect single-chip control solution [80]. The features of TMS320f28335 are:

> High-Performance Static CMOS Technology
> Enhanced Control Peripherals
> Up to 150 MHz (6.67ns Cycle Time)
> IEEE-754 Single-Precision Floating-Point
> Harvard Bus Architecture
> Serial Port Peripherals
> Fast Interrupt Response and Processing
> Code-Efficient (in C/C++ and Assembly)
> On Chip Memory
> 80-ns Conversion Rate
> Advanced Emulation Features
> Real-Time Debug via Hardware
> ANSI C/C++ Compiler/Assembler/Linker
> Code Composer Studio IDE
> Low Power Modes and Power Savings
> Peripheral Interrupt Expansion (PIE) Block
> Flash/OTP/RAM Blocks

Code Composer Studio v4

The development of the controller is carried out using TMS320F28335 which is controlled via an Integrated Development Environment called Code Composer Studio v4 (CCStudio). It is an integrated development environment (IDE) for Texas Instruments' (TI) embedded processor families. A package of tools used to develop and debug embedded applications is collective named as CCStudio. It has compilers for each of TI's device families, source code editor, project build environment, debugger, profiler, simulators, real-time operating system and many other features. CCStudio in combination with Target Support Package software and Real-Time Workshop software, provides an integrated platform that, once installed, requires no coding. In order to load and execute the source code, CCS helps to compile and link the source code so that it can execute on TI DSP [80].

ezDSP Kit

Texas Instruments TMS320F28335 contains a digital signal controller which is evaluated by the developers using a stand-alone card called eZdsp F28335. The main functionality of this module is to develop and run software for the TMS320F28335 processor. It is a complete software development platform for the TMS320F2833x series of floating-point Digital Signal Controllers. The F28335 target board which is part of the starter kit has integrated JTAG emulation, 128Kx16 asynchronous SRAM, CAN 2.0 and RS-232 interfaces, and expansion headers that provide access to all F28335 I/O signals. The Code Composer Studio™ Integrated Development Environment, USB interface to the host PC, and a universal power supply is also part of the starter kit. The stand-alone card eZdsp F28335 allows full speed verification of F28335 code [80].

Fig 5.16. eZdsp F28335 kit [18]

Analog to Digital Converter (ADC) Setup

To enable the designer to interface analog signals directly with the TMS320F28x DSC, 16-channel, 12-bit ADC peripheral is provided. This ADC peripheral included on the TMS320F28xx DSCs has 12 bits of resolution and can achieve a speed of up to 12.5 millions of samples per second (MSPS) (6.25 MSPS and 3.75 MSPS for some parts) through a pipeline architecture that makes it useful in many applications requiring the monitoring of analog signals. Additional features, such as a 16-channel multiplexer, auto-sequencer, dual sample-and-hold (S/H) circuits, and multiple interrupt schemes, make it quite flexible for use in embedded control and data logging applications. The ADC needs to be configured properly to take full advantage of the flexibility of this peripheral. This detailed approach is achieved through various circuit blocks that need to be understood and set up appropriately [80].

The analog input capacity is expanded to 16 channels using two sets (A and B) of 8-channel multiplexers. A single 12-bit ADC core acts as a pipeline analog-to-digital converter. Flexibility of randomly selecting the sequence of ADC input channels connecting to the core is brought by the dual auto-sequencers (SEQ1 and SEQ2). Sequencer1 and Sequencer2 can be cascaded to form a single 16-channel sequencer and each 8-state sequencer can be used independently. ADCMAXCONV register helps in setting the number of conversions per sequence.

ADCRESULT0 – ADCRESULT15, 16 Result Registers to hold the ADC count before they are transferred to system memory are available. To transfer the readings from the result registers to the system memory three different interrupt signals ADCINT, SEQ1INT and SEQ2INT which are generated using the end of sequence. The only CPU intervention to the ongoing ADC operations is the interrupt service routine (ISR) for the transfer of readings from result registers to memory.

In the implementation, cascaded mode is selected. A total of 6 inputs are used. The frequency used is 12.5MHz ADC Star of conversion is triggered by software interrupt. The ADC module will function only with unipolar signals. So it is needed to use offset circuit in the implementation.

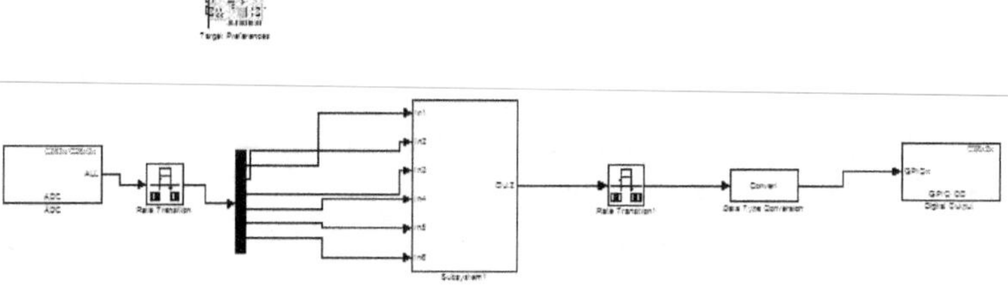

Fig.5.17. Simulink set up for implementation of digital controller

The setup shown in Fig.5.17, consists of a F28335 ADC, the rate transition block, the FVPD algorithm subsystem, Digital output and Target Preference. The Rate Transition Block handles periodic (slow to fast and fast to slow) and asynchronous transitions.

When inserted between two blocks of differing sample rates, The Rate Transition block automatically configures its input and output sample rates for the appropriate type of transition, the user need not specify whether a transition is slow to fast or fast to slow

Fig 5.18 represents the offset circuit used. The offset value is 0.5V. This offset value needs to be subtracted from the sampled input value.

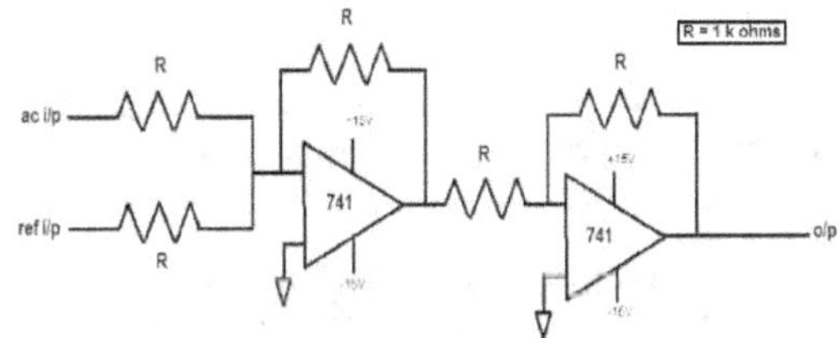

Fig 5.18 Offset circuit

The processor TMS320F28335 is a uni- polar device and the ADC unit of it cannot process the bi-polar values. For this purpose, an offset circuit is designed for voltage input which makes it uni-polar before giving it to the processor. The offset circuit is designed using set of two op-amps which uses the weighted summer principle. The resistors used for the circuit is 1kΩ. One of the inputs given is the ac voltage input obtained from the sensors and the other input is the offset dc voltage that has to be added. For the load that is used, an offset of 0.5V has to be supplied to the voltage offset circuit. The output thus obtained is given to the processor.

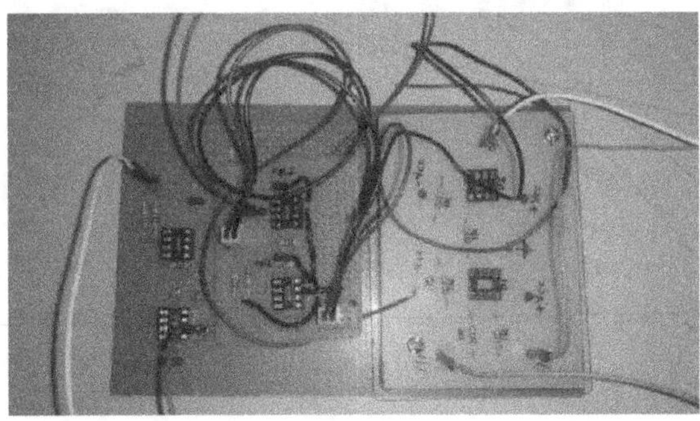

Fig 5.19 Offset circuit

Fig 5.19. shows the offset circuit for the controller, which have two inputs, the ac input and the dc input in order to shift the input voltage before giving it to the kit.

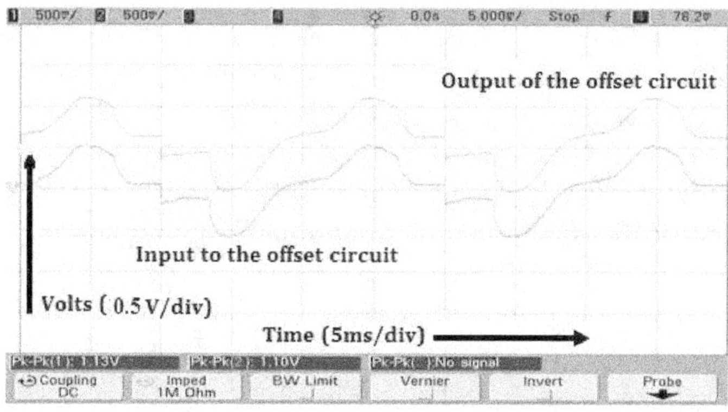

Fig.5.20 Offset input and output of R phase

The Fig.5.20 shows the offset circuit output voltage for R phase input obtained from the voltage sensor.

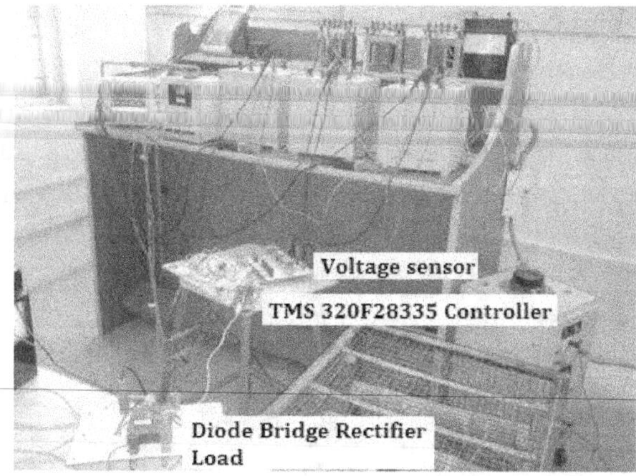

Fig.5.21. hardware setup for the hardware implementation of the controller

Fig.5.21. shows the hardware setup for the hardware implementation of the controller. Diode bridge rectifier with RL load is connected as load. The voltage is reduced to 1V using Voltage sensor and divider circuit. The step down voltage is given to the kit and the pulses are obtained. A three phase source supplying a three phase delta connected R-L load of rating .5kW and 0.1kVAR is considered. The resistance value of load is 70.37Ω and the inductance value is 50mH. The associated distorted voltage waveforms and harmonic table is shown in Fig.5.22.

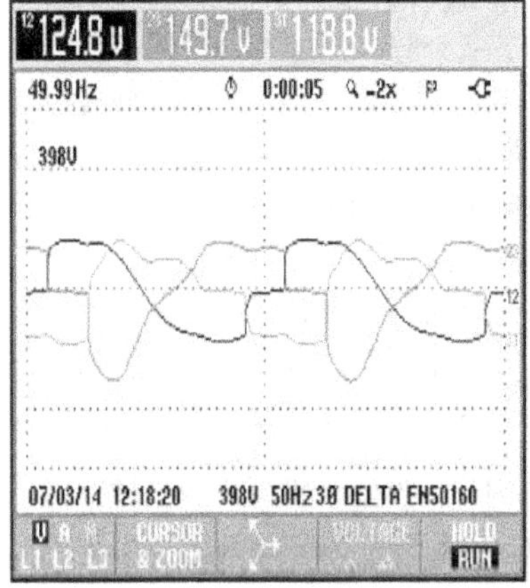

Fig.5.22. Voltage waveforms and harmonic table

For the hardware setup shown in Fig.5.21, the sensed voltage is first given to the offset circuit and then to TMS320F28335 controller and output pulses shown in Fig.5.23 are taken from GPIO port.

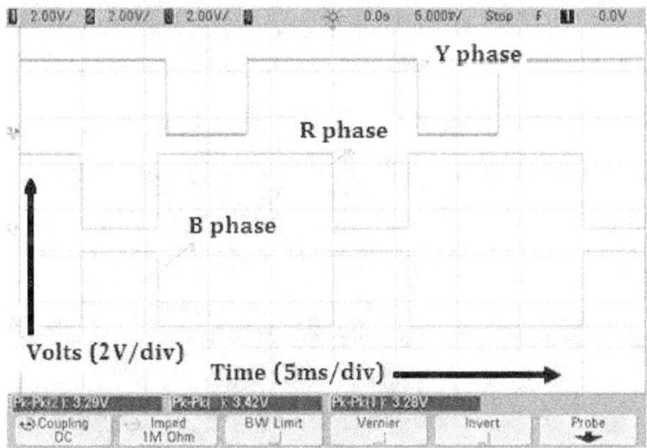

Fig.5.23. Pulses obtained with digital controller

One of the drawbacks of linking Simulink circuit with controller and obtaining the pulses is that there is no possibility of viewing the intermediate outputs. So step wise analysis is difficult whereas this is possible in analog implementation.

5.6.2. Analog Implementation of the controller

A diode bridge rectifier connected to three phase, 415V, 50 Hz supply is the hardware system configuration considered for hardware implementation. 50% of the rated voltage is used for testing. The output of diode bridge rectifier is connected to a lamp load of 1kW which contains ten numbers of 100W bulbs each. Three phase 415 V,50Hz supply is connected to diode bridge rectifier via source inductance of 10mH and isolation transformer of 2kVA rating with 50% tapping so that the input voltage to DBR is 110V(phase). The total rating of DBR(36MT160) with lamp load is 300W, which is considered for analysis of harmonic condition. The diode bridge rectifier draws highly nonlinear currents which in turn distorts the voltage. This distorted voltage is sensed using voltage sensor (EM010TEMPHIN by ABB) circuit and given as input to the controller. All the components in the block diagram representation of the FVPD controller has been designed with corresponding analog circuit components and output of the controller has been measured. The key components used are LM 741 or LM 324, AD 711 and AD633.These components were assembled to form a PCB prototype of FVPD controller.

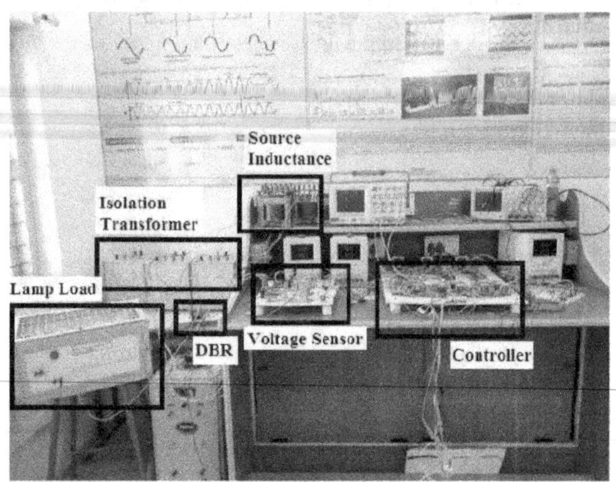

Fig.5.24. Hardware Setup

The voltage sensor used for the implementation of the proposed system is from ABB and the series is EM010TEMPHIN. It is based on Closed Loop Hall Effect Technology. The voltage sensor converts 110V to 1V.Fig.5.25 and Fig.5.26 represent the image and testing circuit of voltage sensor.

Fig.5.25. Image of Voltage sensor

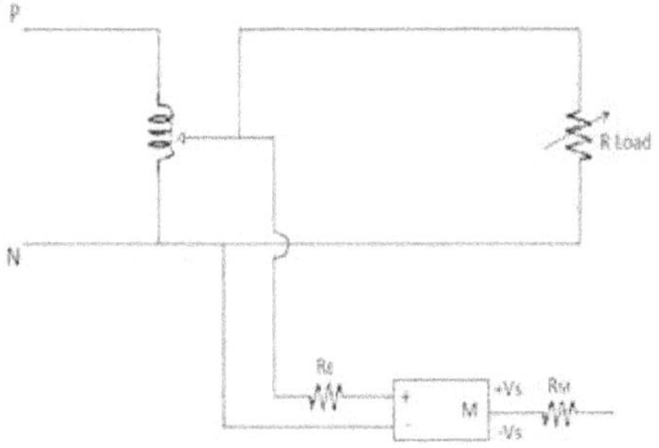

Fig.5.26. Circuit of Voltage sensor

5.6.2.1. Testing of Control Circuit Components

The main components in analog controller are fundamental detector circuit, phase shifter, peak detector, divider circuit, multiplier circuit and comparator etc. Each component is represented with analog circuit and testing analysis has been done. The controller circuit is shown in Fig.5.27.

Fig.5.27. Controller Circuit

a) *Fundamental Detector Circuit/Low pass filter:*

A second order biquad filter with cut off frequency 50Hz is used as the fundamental voltage detector circuit to extract the fundamental frequency component. The circuit made using 741 IC is shown in Fig.5.28.

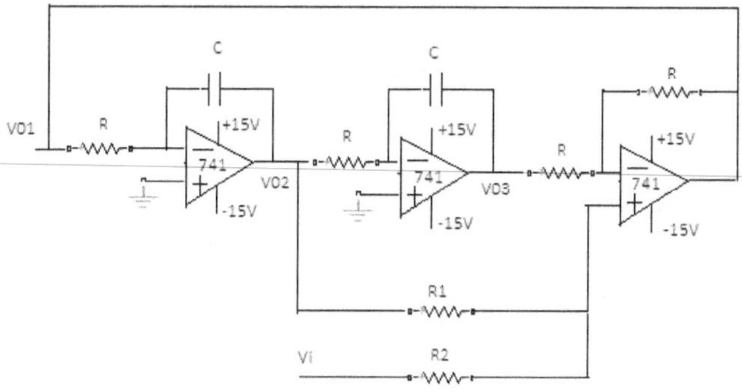

Fig.5.28. Analog circuit of low pass filter.

Where R = 32kΩ, C = 0.1μF, R_1 = (1-α) *R = 16 kΩ, R_2 = α*R = 16 kΩ,

V_{01} = high pass filter output,

V_{02} = band pass filter output,

V_{03} = low pass filter output and

V_i = input.

The low pass equation is

$$V_{03} = \frac{2(1-\alpha)}{\left[\frac{s^2}{\omega_0{}^2}+\frac{s}{\omega_0*Q}+1\right]}$$

(5.29)

Where cut-off frequency ω_0 = 1/RC, Q = 1/2α

The gain of the low pass filter is 2 (1-α). To make the gain as unity, α is selected as 0.5. The output of the biquad filter hence is of fundamental frequency, 90^0 phase shifted and the same magnitude as that of the input.

Fig.5.29 represents the input and output voltage waveforms of low pass filter.

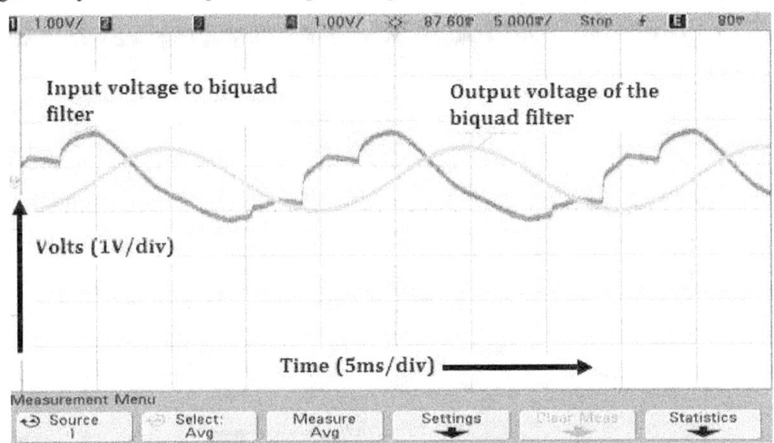

Fig.5.29   Input and output voltage waveforms of low pass filter

The sensed voltage from the voltage sensor is given as input to low pass filter which extracts the fundamental component with an inherent 90^0 phase shift as shown in Fig.5.29.

b) *Phase shifter*:

The phase shifter is used to remove the 90^0 phase shift created by the biquad filter. It is an all pass filter as shown in Fig.5.30. The transfer function is given as

$$H(s) = \frac{s-\omega_0}{s+\omega_0}$$

(5.30)

Where $\omega_0=1/RC$ and the expression for phase angle is given as,

$$\theta = 180^0 - 2tan^{-1}\frac{\omega}{\omega_0} \tag{5.31}$$

At $\omega = \omega_0$, θ=90^0.The value of R and C to get 90^0 phase shift are given as R=32kΩ and C=0.1µF

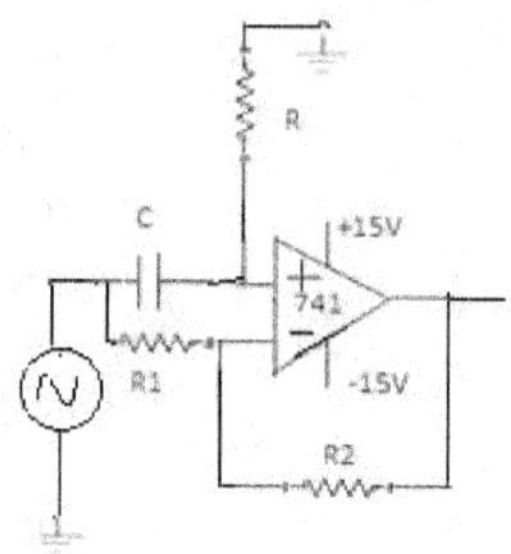

Fig.5.30. Phase shifter Circuit

Where R$_1$=R$_2$=10 kΩ, R=32 kΩ, and C=0.1µF

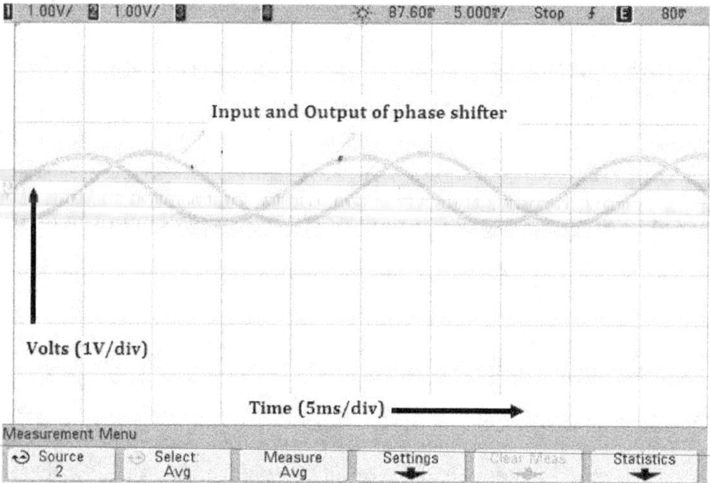

Fig.5.31. Phase shifter input and output

Fig.5.31 represents phase shifter input and output. It is seen that the 90^0 phase shift is created by the phase shifter circuit. The inherent 90^0 lag created by the biquad filter can be removed by connecting phase shifter which provides 90^0 lead to cancel the lagging phase shift, so that the output of the phase shifter will be a signal which is in phase with the sensed input voltage which clearly shown in Fig.5.32

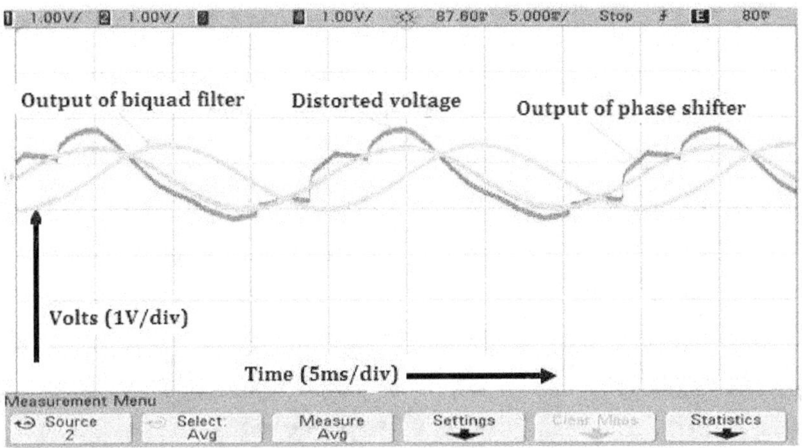

Fig.5.32. Sensed voltage, low pass filter output and phase shifter output

c) *Peak Detector*:

Peak detector is basically a rectifier circuit with capacitive filter. The 0.7 V which is the cut-in voltage of diode can be compensated for, by connecting an adder in the output of the rectifier. The circuit diagram is shown in Fig.5.33 and corresponding waveforms are shown in Fig.5.34.

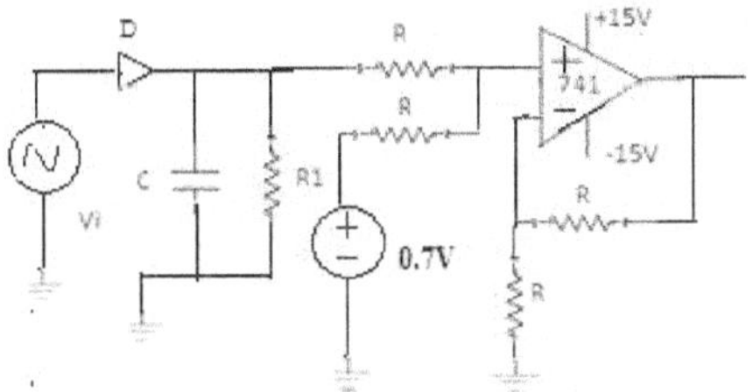

Fig.5.33. Peak detector circuit

Where R_1=1K Ω, R=330Ω, C=470μF

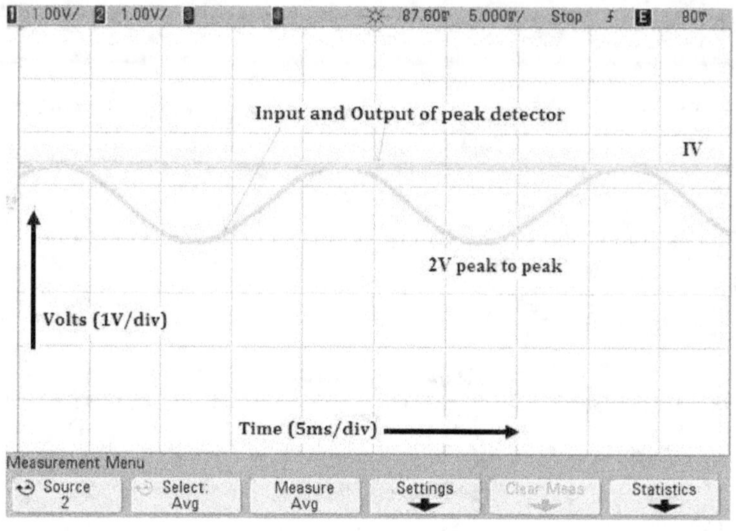

Fig.5.34. Input and Output of Peak detector circuit

d) Divider Circuit:

Fig.5.35 represents the divider circuit. A multiplier is connected in feedback to get the divider circuit and transfer function is given by $W' = -10(E/E_x)$ (5.32)

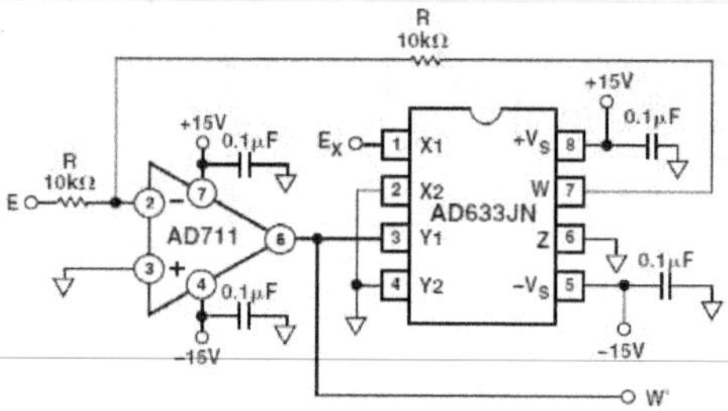

Fig.5.35. Divider Circuit

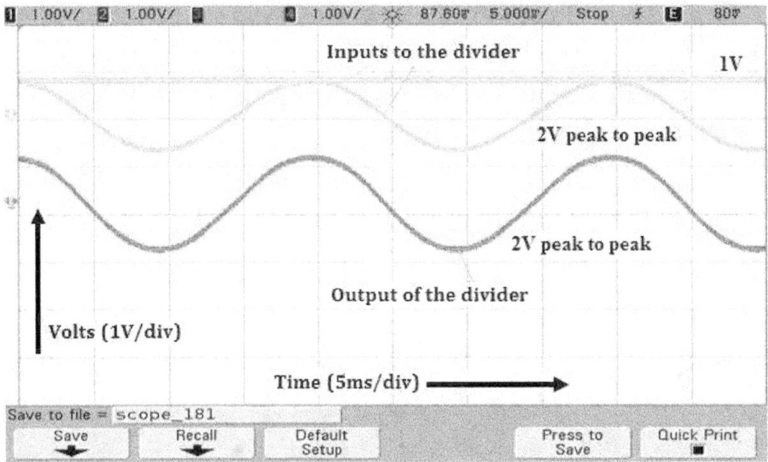

Fig.5.36. Inputs and output of the Divider Circuit

The output of the peak detector and output of phase shifter is given as inputs to the divider circuit to get unit amplitude sine wave as output which is depicted in Fig.5.36. The divider circuit output is inverting with a gain of 10 and to compensate that an inverting amplifier of gain 0.1 is connected at divider output.

e) Multiplier Circuit:

AD633 IC is used as multiplier to multiply unit amplitude sine wave with rated magnitude so as to generate reference source voltage which is the expected signal at PCC.

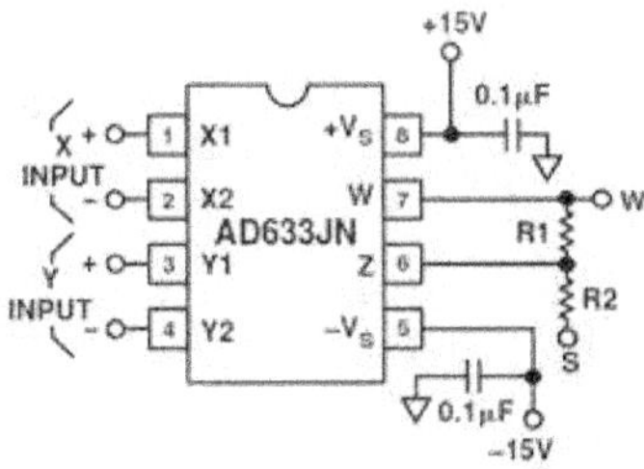

Fig.5.37. Multiplier Circuit

$$W = \frac{(X_1 - X_2)(Y_1 - Y_2)}{10} \times \frac{(R_1 + R_2)}{R_1} + S \qquad (5.33)$$

Where S is the summing input whose value is taken as zero since there is no additional signal added. The values of R_1 and R_2 are chosen in such a way that the overall gain becomes unity.

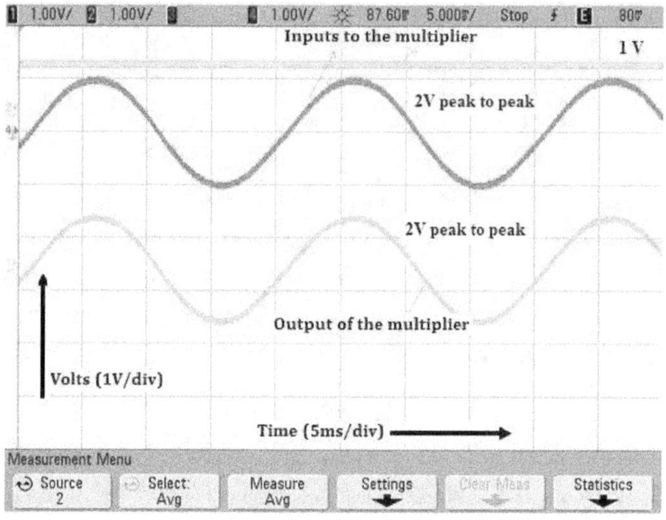

Fig.5.38. Inputs and output of the Multiplier Circuit

The output of the multiplier as shown in Fig.5.38 represents the reference voltage at PCC. The inputs are also represented in the diagram which are the output of divider circuit (unit amplitude sine wave) and the rated voltage magnitude (in this case it is 1V).

f) Subtractor and pulse generator(Comparator) unit:

Once reference signal at PCC is obtained, reference filter voltage can be calculated by subtracting sensed source voltage from reference voltage at PCC. The corresponding circuit diagram is shown in Fig.5.39. The output of the subtractor $V_{fa} = V_{saref} - V_{sa}$. Both inputs and output waveforms are illustrated in Fig.5.40.

The reference filter voltage is compared with actual filter voltage to generate pulses to the inverter which is also depicted in Fig.5.39.

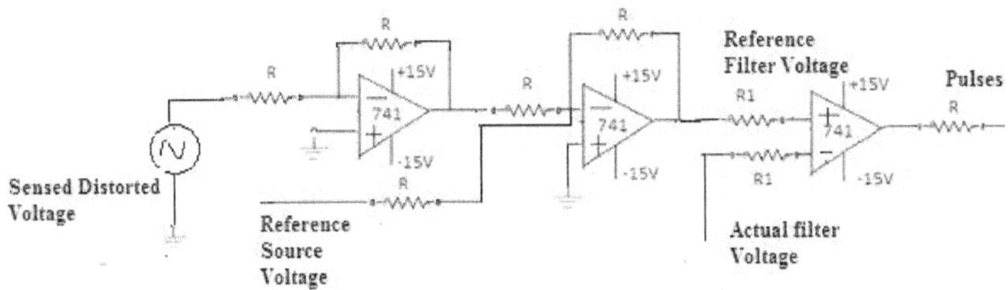

Fig.5.39. Circuit diagram for Subtractor and pulse generation circuit

Where R=10 kΩ, R₁ =1kΩ.

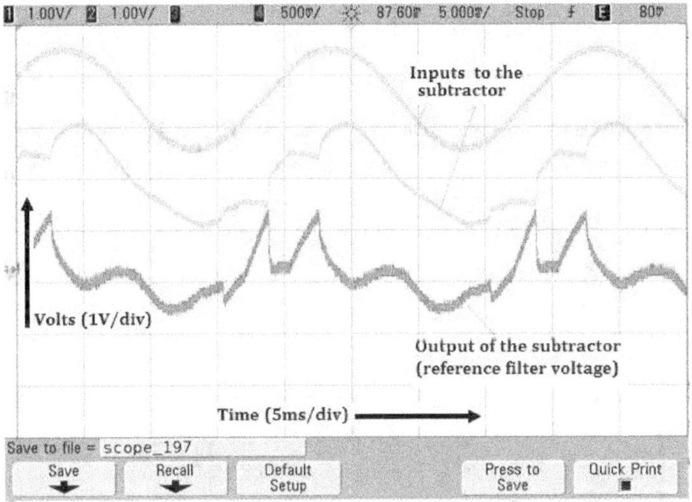

Fig.5.40. Inputs and output of the Subtractor Circuit

Fig.5.41 represents the waveforms associated with the pulse generator (comparator) circuit. The input signals are reference and actual filter voltages. For testing, the actual filter voltage is considered as ground. It is seen that when the reference filter voltage is higher than the actual filter voltage, the output of the comparator is high, else it is low.

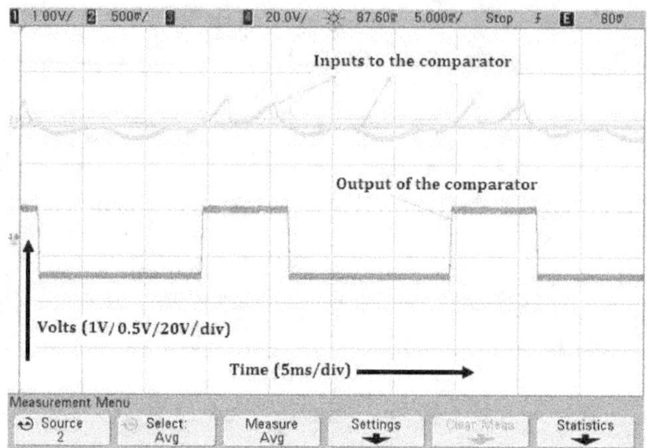

Fig.5.41. Inputs and output of the Comparator Circuit

The same comparator circuit by interchanging the inputs is implemented to generate the complementary pulses and the corresponding waveforms are shown in Fig.5.42.

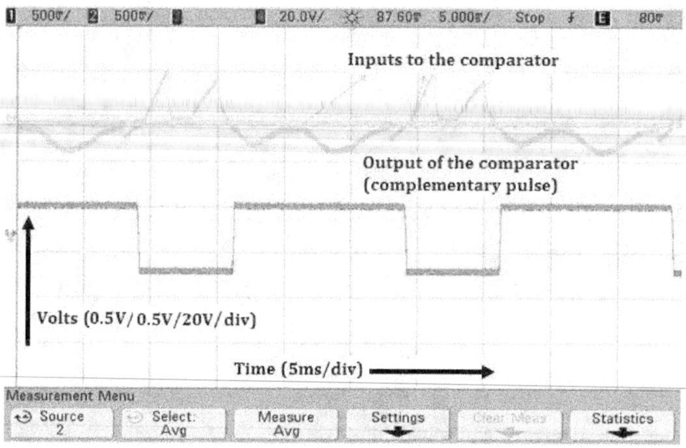

Fig.5.42. Inputs and output of the Comparator Circuit (Complementary pulse)

If there is any phase difference, then to do phase correction so as to maintain 120^0 phase difference between each phase, before feeding the output signal of the divider circuit to multiplier circuit, phase-shifting can be done to generate unit vector templates by taking 'A' phase as reference and then multiplier circuit is connected to get reference source voltages. The overall controller circuit is shown in Fig.5.43.

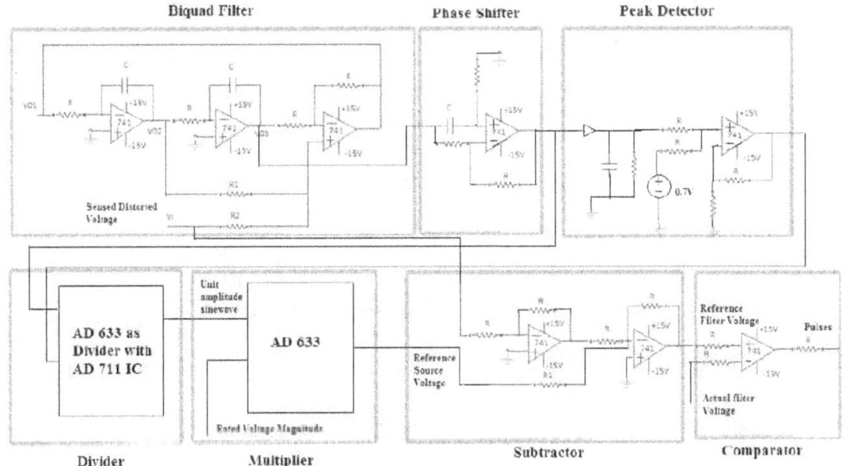

Fig.5.43. Analog controller Circuit

5.6.2. 2.Performance Analysis of the proposed FVPD controller:

5.6.2.2.1 Under harmonic Conditions:

A 300W diode bridge rectifier connected to three phase, 50 Hz supply is considered for hardware implementation of the system. The diode bridge rectifier is fed via three star connected isolation transformer whose primary side is connected to the supply through a series inductor. The highly nonlinear current drawn by the diode bridge rectifier induces voltage harmonics into the system. These distorted source voltages are sensed by the voltage sensor and given to the analog controller. The analog controller determines the reference filter voltages that has to be injected by the series active filter. Fig.5.44 (a) represents the three phase distorted source voltages. In Fig.5.44(b) the source voltage, reference filter voltage and output voltage after series compensation which is nothing but the expected voltage at load terminals are shown (this is the voltage at the load terminal obtained when the controller reference output voltage is added with the sensed PCC voltage, to verify the effectiveness of the controller under harmonics conditions). Fig.5.44(c) and Fig.5.44 (d) show the three phase reference filter voltages and output after series compensation, shown here as sum of reference voltage and sensed PCC voltage, as mentioned above respectively.

From the waveforms, it is seen that the distorted voltage is changed to pure sinusoidal waveform by the elimination of harmonic components with the help of the Series Active Filter.

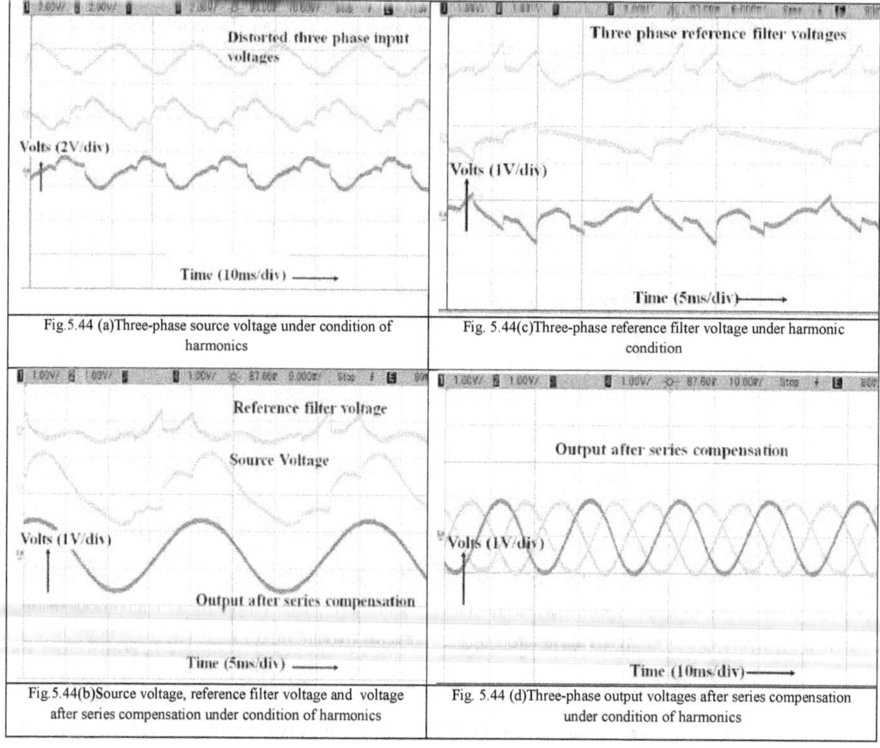

Fig.5.44 (a)Three-phase source voltage under condition of harmonics	Fig. 5.44(c)Three-phase reference filter voltage under harmonic condition
Fig.5.44(b)Source voltage, reference filter voltage and voltage after series compensation under condition of harmonics	Fig. 5.44 (d)Three-phase output voltages after series compensation under condition of harmonics

The above figures(Fig.5.44) are obtained from the hardware implementation of the proposed FVPD controller which indicates that a pure sinusoidal wave is obtained by the controller action where the input is distorted wave.

5.6.2. 2.2. *Under three phase voltage Sag and Swell*:

For testing the cases like sag and swell, three phase, AC 415 V, 50 Hz supply is connected to a 100 W lamp load, through autotransformers without source inductance. In this case, measurement resistor at the sensor terminals is designed such that it converts 110 V to 1 V peak amplitude. Above and below 110 V is considered as swell and sag respectively, for testing the controller prototype. A three phase resistive load is considered as the load for the analysis. Similar waveforms as before are obtained and recorded in the case of sag and swell also.Fig.5.45 and

Fig.5.46 represents the waveforms obtained using three phase sag and swell respectively. From the waveforms it is seen that the injected voltage in the case of sag and swell are in phase and 180^0 out of phase with respect to system voltage respectively. The in phase component boosts up the voltage level where as the 180^0 out of phase component decreases the value of voltage once it is added so that in both sag and swell cases the voltage at the PCC can be restrained to rated voltage. This control action happens almost instantaneously so that   even   the sensitive loads connected will not adversely get affected because of the voltage changes. The results prove the effectiveness of the controller for restoration of rated voltage during three phase sag and swell conditions.

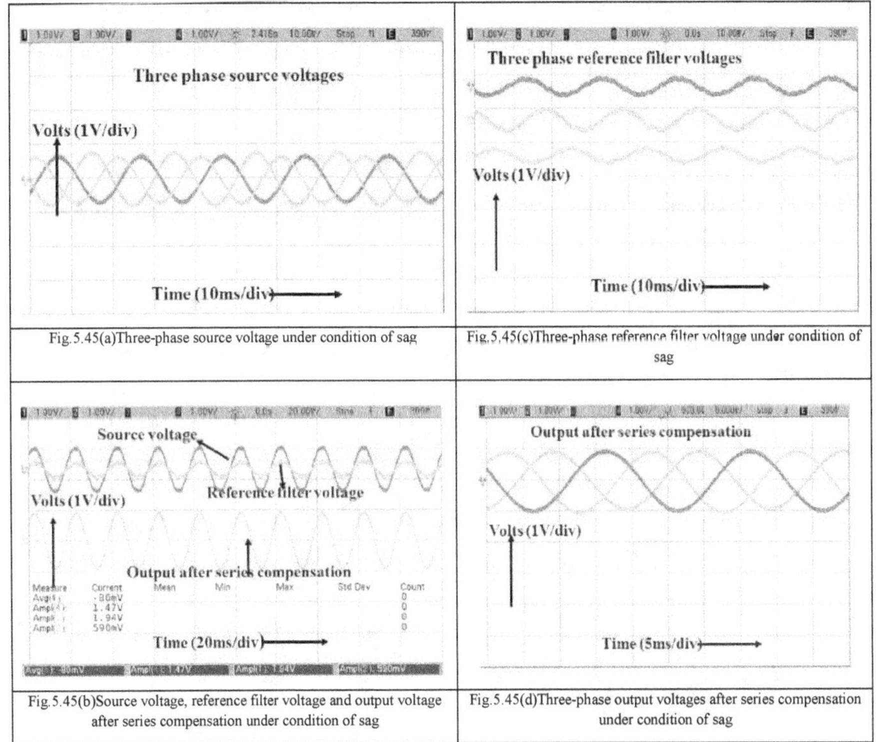

Fig.5.45(a)Three-phase source voltage under condition of sag	Fig.5.45(c)Three-phase reference filter voltage under condition of sag
Fig.5.45(b)Source voltage, reference filter voltage and output voltage after series compensation under condition of sag	Fig.5.45(d)Three-phase output voltages after series compensation under condition of sag

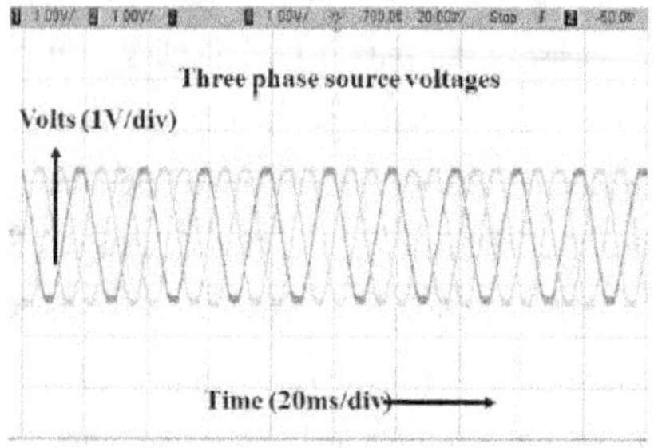

Fig.5.46(a)Three-phase source voltage under condition of swell	Fig.5.46(c)Three-phase reference filter voltage under condition of swell
Fig.5.46(b)Source voltage, reference filter voltage and output voltage after series compensation under condition of swell	Fig.5.46(d)Three-phase output voltages after series compensation under condition of swell

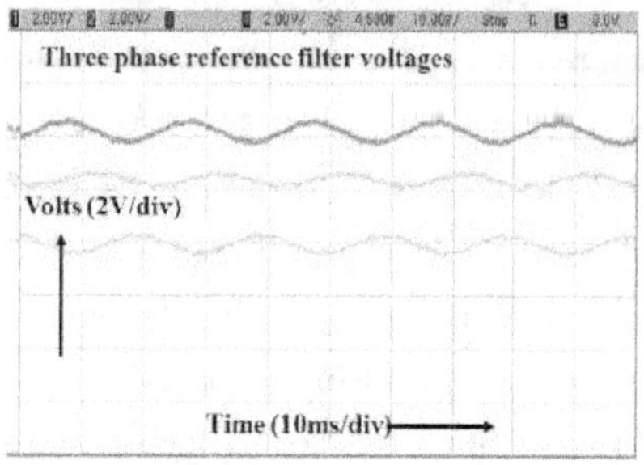

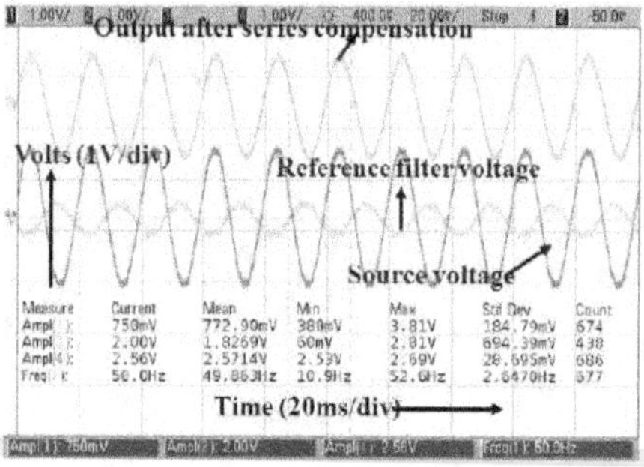

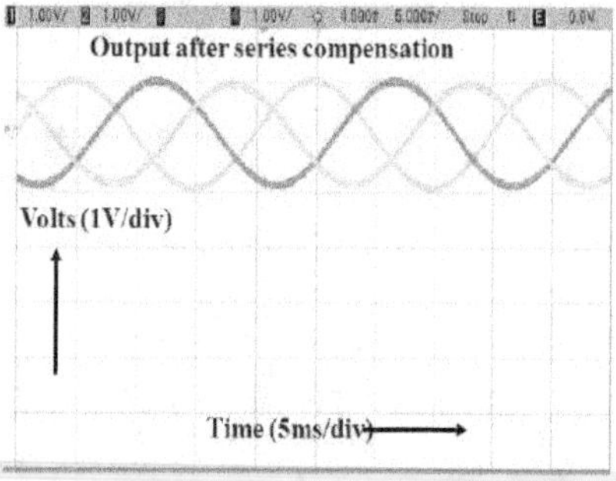

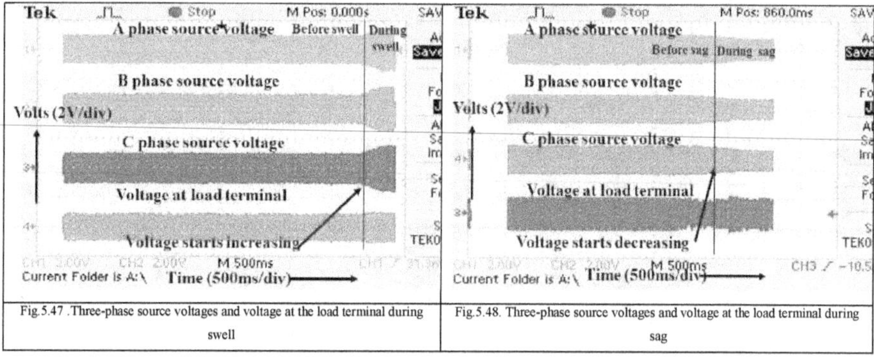

Fig.5.47 .Three-phase source voltages and voltage at the load terminal during swell	Fig.5.48. Three-phase source voltages and voltage at the load terminal during sag

Fig.5.47 and Fig.5.48 represent the three-phase source voltages, and voltage at load terminal (output after series compensation), in the case of three-phase swell and sag respectively. Even though sag and swell occur, voltage at the load terminal remains unaltered, due to the corrective actions performed by the proposed FVPD controller. The FVPD controller injects voltage in phase with the system voltage for compensating sag and injects out of phase with system voltage in the case of swell. The point at which sag/swell occurs is also indicated in the figure. Before and after occurrence of sag/swell the load voltage remains the same.

5.6.2.2.3. Under Magnitude Unbalance:

To analyze the performance of the controller under magnitude unbalance, the magnitude of one of the phases is reduced compared to other two phases which are at rated voltages and given to the controller. The associated waveforms are shown in Fig.5.49. The figures indicate that compensation voltage is generated only for the phase where reduction is occurred.

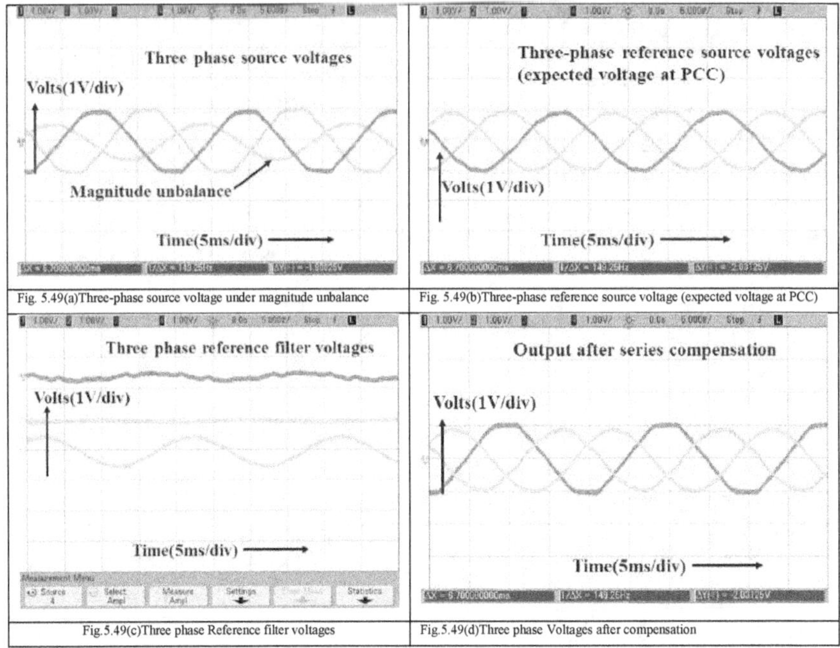

Fig. 5.49(a)Three-phase source voltage under magnitude unbalance	Fig. 5.49(b)Three-phase reference source voltage (expected voltage at PCC)
Fig.5.49(c)Three phase Reference filter voltages	Fig.5.49(d)Three phase Voltages after compensation

Fig. 5.49(a) shows three phase source voltages with magnitude unbalance. In Fig. 5.49(b) reference source voltages are shown. Three phase reference filter voltages (compensation voltages) and source voltages after compensation are represented in Fig.5.49(c) and Fig.5.49(d)respectively. From Fig.5.49(c) it is observed that the compensation signals are generated only for that particular phase whose magnitude is not equal to rated value. In the other two phases, the compensating voltages are zero. After the compensation all the three phases are restored to their rated value with the timely action of proposed FVPD controller.

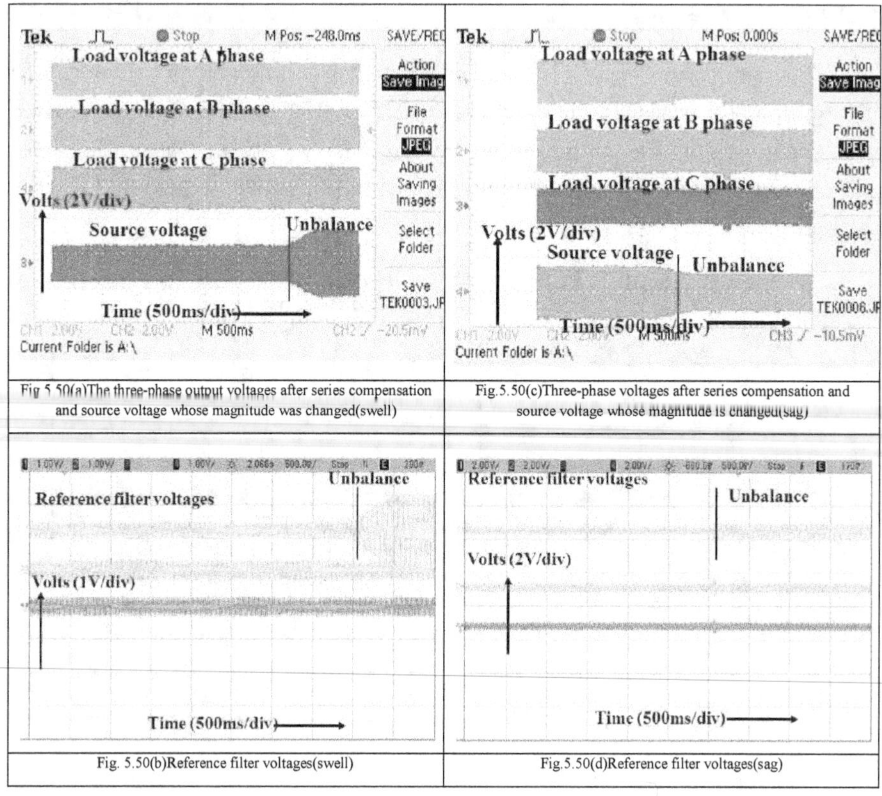

Fig 5.50(a)The three-phase output voltages after series compensation and source voltage whose magnitude was changed(swell)	Fig.5.50(c)Three-phase voltages after series compensation and source voltage whose magnitude is changed(sag)
Fig. 5.50(b)Reference filter voltages(swell)	Fig.5.50(d)Reference filter voltages(sag)

Fig.5.50 also illustrates that the magnitude imbalance, sag or swell was introduced in one phase only. Fig. 5.50(a). shows the three-phase output voltage after series compensation. It also displays source voltage whose magnitude is changed (Fig. 5.50(a)) and reference filter voltages. (Fig. 5.50(b)). The reference filter voltage, of only that unbalanced phase, is generated to carry out

the compensation. Fig. 5.50(c), (d) also represents similar waveforms during unbalanced sag condition; the corresponding reference filter voltage is increased, to maintain the rated voltage at load terminals. Only when unbalance is there a compensation voltage is generated by the FVPD controller. This indicates that the controller works well in unbalanced condition (increase or decrease in magnitude)

5.6.2.2.4. Under Phase Unbalance:

Performance of the proposed FVPD controller has been analyzed by introducing a phase difference other than 120^0 and 240^0 by changing phase of one of the signals which are depicted in Fig.5.51.

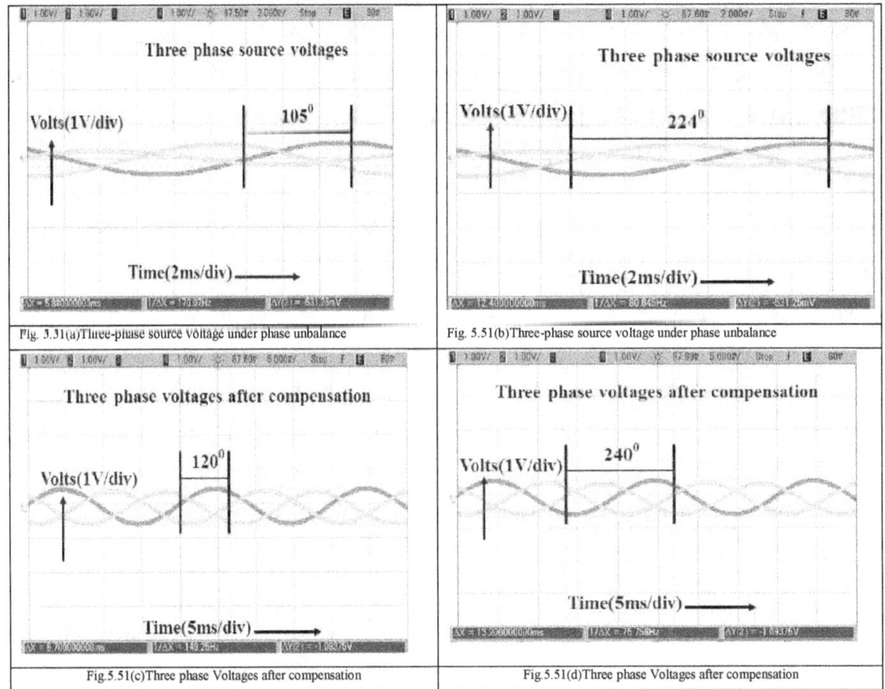

Fig. 5.51(a)Three-phase source voltage under phase unbalance	Fig. 5.51(b)Three-phase source voltage under phase unbalance
Fig.5.51(c)Three phase Voltages after compensation	Fig.5.51(d)Three phase Voltages after compensation

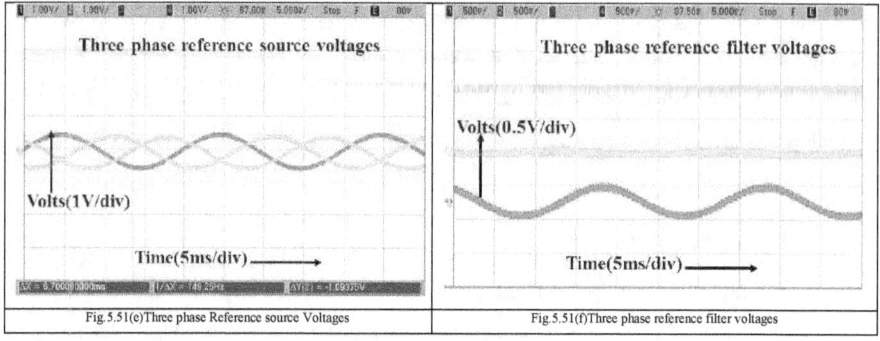

Fig.5.51(e)Three phase Reference source Voltages	Fig.5.51(f)Three phase reference filter voltages

Fig.5.51. (a) and (b) represent the source voltage waveforms with phase unbalance. In Fig. 5.51(a) its seen that the time difference between two signals are 5.88ms instead of 6.7ms, which means the phase difference between them is 105^0 instead of usual 120^0. Similarly, from Fig. 5.51(b), the phase difference between the signals are 225^0 corresponds to 12.4ms instead of 240^0. It is observed from Fig.5.51(c) and (d), which represents the voltage after compensation that the time difference between the signals became 6.7ms and 13.2ms corresponding to 120^0 and 240^0 respectively as required. This phase correction has been done by the controller. Fig.5.51 (e) and (f) represent three phase reference source voltage and compensating voltage respectively. Compensating voltage is generated for only one phase where the phase difference happened. For the other two phases the reference filter voltages are zero. Phase correction is also carried out where phase unbalance has occurred. From the analysis carried out under conditions like voltage harmonics, three phase sag and swell, magnitude and phase unbalance the proposed FVPD controller was found to be very effective in compensating all kinds of voltage perturbations. The controller requires very less components in hardware implementation thereby effectively reducing the cost and memory.

5.6.3. Response time of proposed FVPD controller

During simulation analysis, the response time of the proposed controller was less than half a cycle. This was verified empirically, by, introduction of sudden increase or decrease in source voltage, as shown in Fig.5.52 (a) and (b) respectively. The benign feature of controller's near-instantaneous response was self-evident, when the filter's reference voltage also increased, with each change in the source voltage. From Fig.5.52, it is also seen that the reference voltage is in phase with source voltage in the case of sag and out of phase with the system voltage in the case of swell respectively.

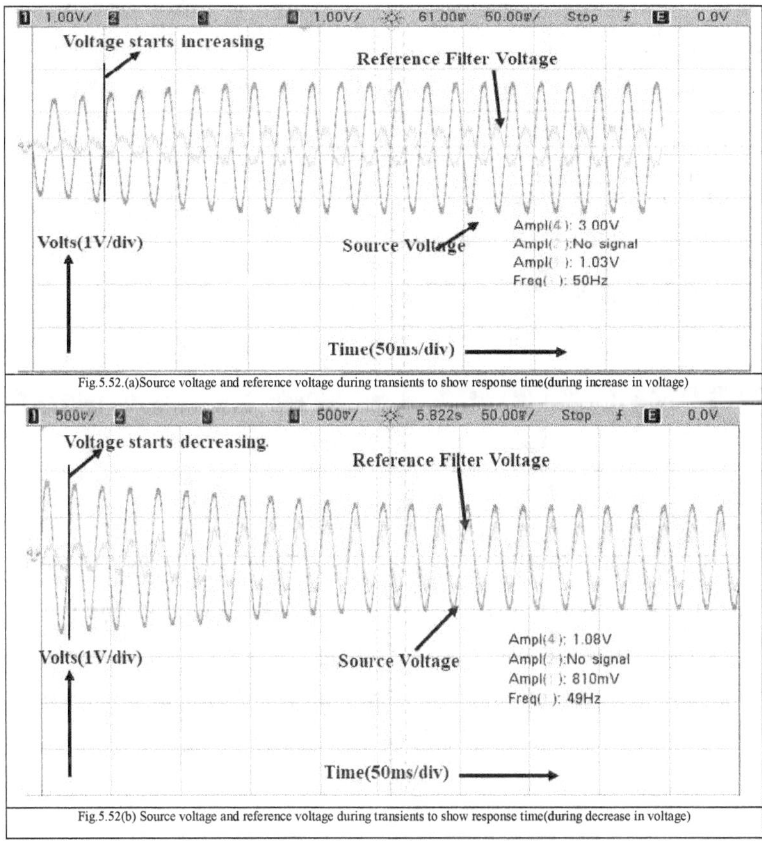

Fig.5.52.(a)Source voltage and reference voltage during transients to show response time(during increase in voltage)

Fig.5.52(b) Source voltage and reference voltage during transients to show response time(during decrease in voltage)

5.7 Performance Comparison of proposed FVPD control Algorithm with IRPT and SD Control Algorithms

To check the feasibility of the proposed FVPD controller, its performance is compared with already existing controllers based on IRPT and SRF algorithms which are explained in section 5.2.

For harmonic compensation, the three-phase 400V, 50Hz, programmable source, connected to three-phase, 5kW resistive load is considered as the test system. Harmonics are introduced into the voltage with the help of programmable voltage source during the period 0.05 to 0.1 secs to carry out the simulation analysis. For compensation of sag and swell, three-phase source, with RL load, considered with the active and reactive powers of load as 5kW and 100VAr, respectively. The simulation parameters are given in Table 5.4. In the simulation exercises, different cases of voltage perturbations were considered.

Table 5.4. Simulation Parameters

Load	5kW,100VAr
kVA rating of Inverter	4.5 kVA
kVA rating of Transformer	4.5 kVA
Vdc	650V
DC link capacitor	9mF
Interfacing Inductor	25mH

Case 1: Compensation of Voltage Sag and Swell

In this case, three phase voltage sag of 20% was introduced in the source voltage and simulations conducted, with IRPT based controller, SRF based controller, followed by the proposed FVPD controller. Fig.5.53 presents the source voltage, reference filter voltage, and voltage at the load terminals, with these three controllers. Test observations show that the proposed controller works well; it is possible to restore the rated voltage at load terminals, with the FVPD controller, whereas the compensations are incomplete with IRPT and SRF controllers.

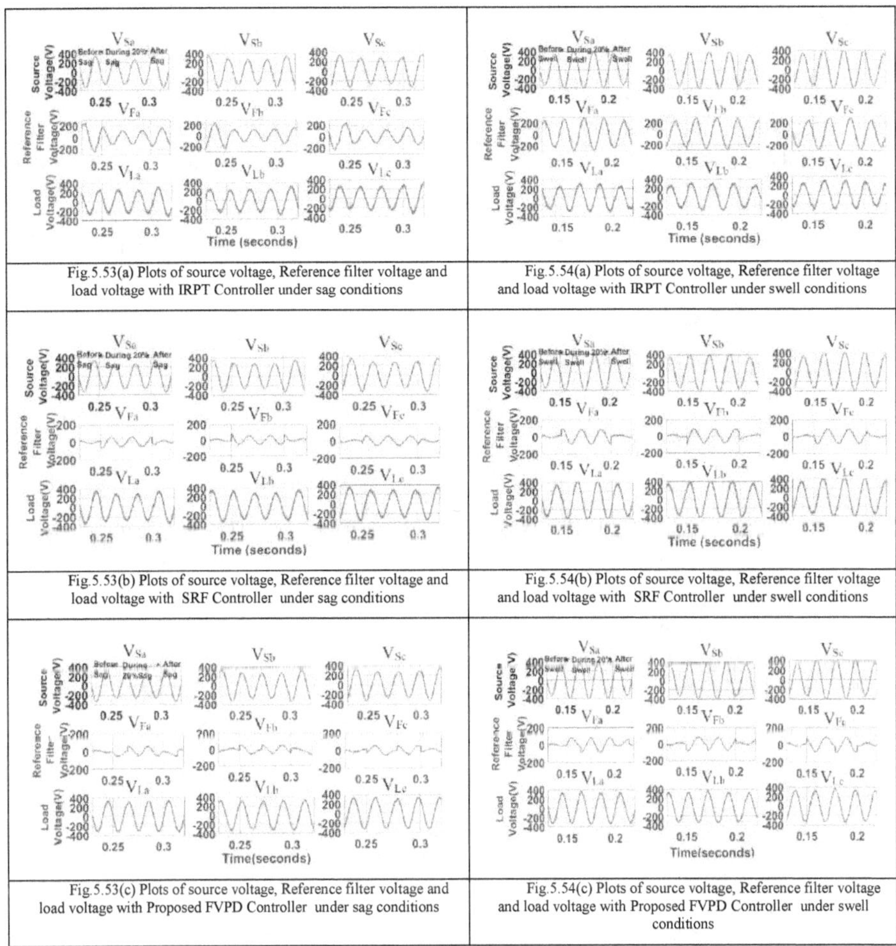

Fig.5.53(a) Plots of source voltage, Reference filter voltage and load voltage with IRPT Controller under sag conditions	Fig.5.54(a) Plots of source voltage, Reference filter voltage and load voltage with IRPT Controller under swell conditions
Fig.5.53(b) Plots of source voltage, Reference filter voltage and load voltage with SRF Controller under sag conditions	Fig.5.54(b) Plots of source voltage, Reference filter voltage and load voltage with SRF Controller under swell conditions
Fig.5.53(c) Plots of source voltage, Reference filter voltage and load voltage with Proposed FVPD Controller under sag conditions	Fig.5.54(c) Plots of source voltage, Reference filter voltage and load voltage with Proposed FVPD Controller under swell conditions

Fig.5.54 represents the source voltage, reference filter voltage, and voltage at load terminals, with IRPT, SRF and proposed FVPD controllers under conditions of 20% voltage swell. Compared to the other two, the performance of the proposed FVPD controller is much better, as voltage at the load terminal is restored to the rated voltage by the proposed controller, whereas with other two controllers, full compensation of load voltage could not be achieved. In Fig.5.53 and Fig.5.54, three conditions such as before sag, during sag and after sag is shown. The reference voltage is generated only during sag condition with proposed FVPD controller whereas with IRPT and SRF controllers, even before and after sag conditions also compensating voltages are generated.

Case 2: Compensation of harmonics:

In this case, fifth and seventh order harmonic voltages were induced in the source voltage using programmable voltage source to test the performance of the controller. From Fig.5.55, it is seen that the source voltage is distorted, and with the FVPD controller action, it is possible to obtain sinusoidal voltage at load terminals after compensation.

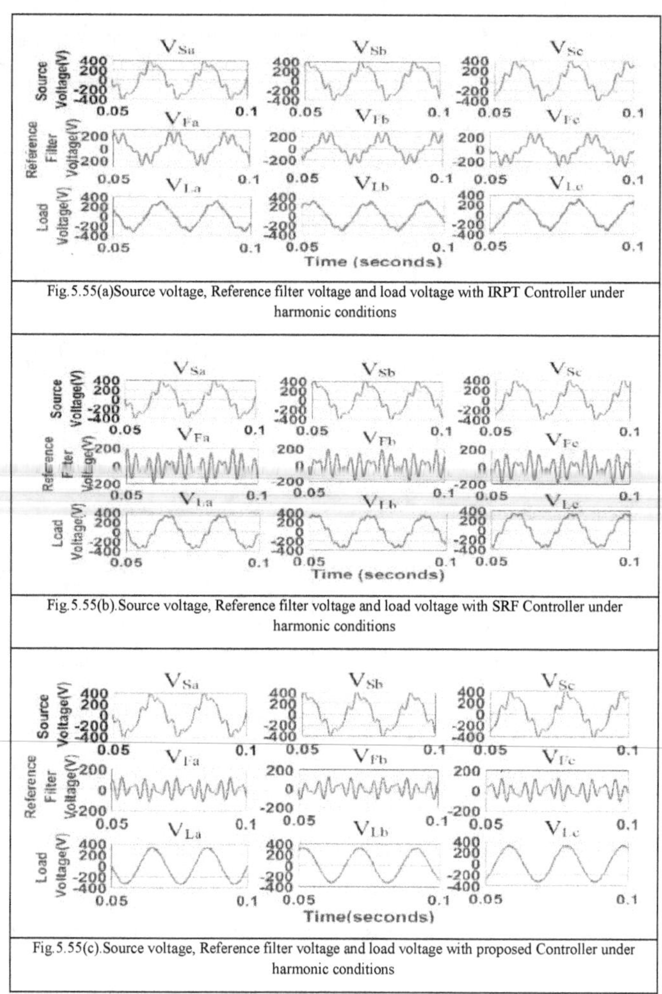

Fig.5.55(a)Source voltage, Reference filter voltage and load voltage with IRPT Controller under harmonic conditions

Fig.5.55(b).Source voltage, Reference filter voltage and load voltage with SRF Controller under harmonic conditions

Fig.5.55(c).Source voltage, Reference filter voltage and load voltage with proposed Controller under harmonic conditions

The THD before and after compensation is shown in Table 5.5

Table 5.5: THD Analysis of Source Voltage

%THD	Proposed FVPD	IRPT	SRF
before compensation	25%	25%	25%
after compensation	3%	10.43%	12.9%

The THD in source voltage is reduced from 25% to 3% with proposed FVPD controller which is well within the limits prescribed by IEEE 519 standard whereas with IRPT and SRF based controllers the THD after compensation is 10.43% and 12.9% respectively.

Case 3: Compensation of Voltage Magnitude Unbalance:

Magnitude unbalance is created by introducing 20% voltage sag in phase 'A' only. The associated waveforms are shown in Fig.5.56. It is seen that with proposed FVPD controller the voltage at all the three phases are restored back to the normal specified voltage whereas the complete compensation was not achieved with IRPT or SRF based controllers.

Case 4: Compensation of Phase Unbalance:

To create phase imbalance, additional phase delay of 20^0 is introduced between two phases, so that the phase difference between 'A' and 'B' phases is 100^0, and phase difference between 'A' and 'C' is 220^0. After compensation, the proposed FVPD controller retains the phase difference as 120^0 and 240^0, respectively; however, with the other two controllers, the phase difference is not exactly 120^0 as observed in Fig.5.57. It is observed that with IRPT and SRF based controllers the complete phase correction was not able to perform. In the diagram the mentioned load voltage is nothing but the voltage at PCC where the loads are actually connected to the system. The results indicate that the proposed FVPD controller achieves complete phase correction compares to IRPT and SRF based controllers.

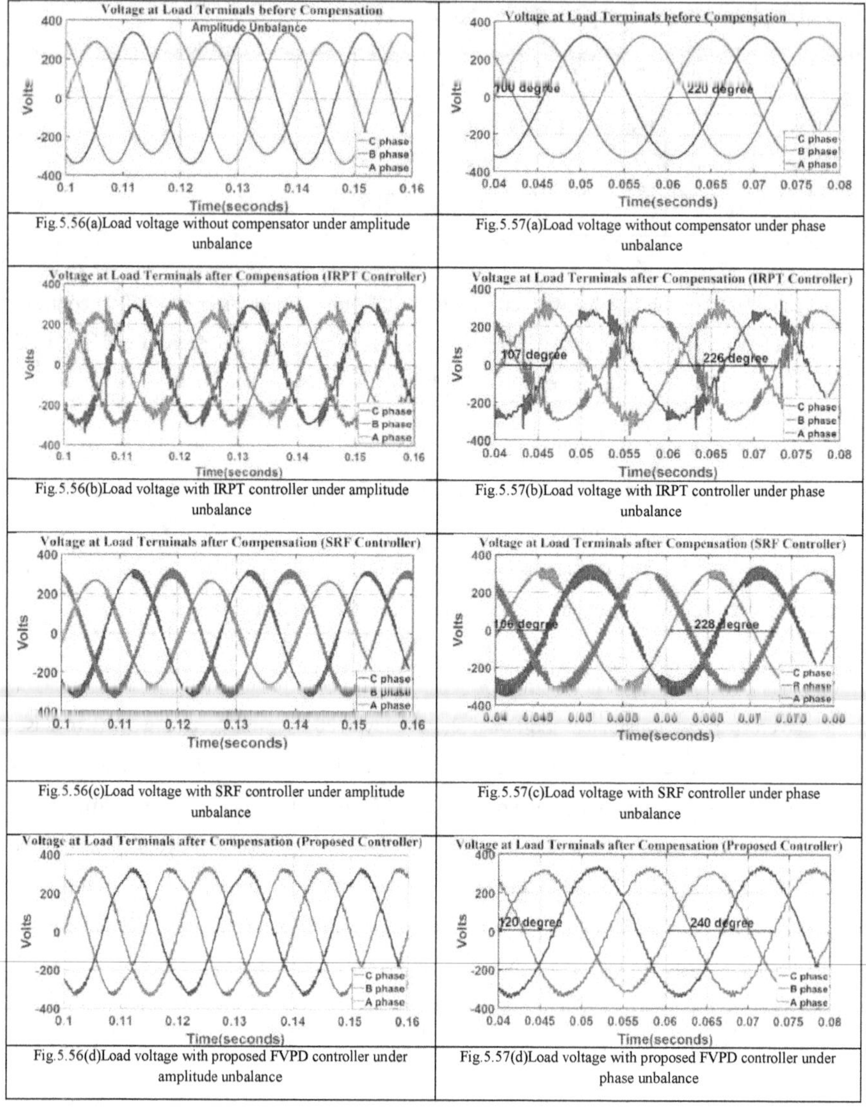

Fig.5.56(a)Load voltage without compensator under amplitude unbalance	Fig.5.57(a)Load voltage without compensator under phase unbalance
Fig.5.56(b)Load voltage with IRPT controller under amplitude unbalance	Fig.5.57(b)Load voltage with IRPT controller under phase unbalance
Fig.5.56(c)Load voltage with SRF controller under amplitude unbalance	Fig.5.57(c)Load voltage with SRF controller under phase unbalance
Fig.5.56(d)Load voltage with proposed FVPD controller under amplitude unbalance	Fig.5.57(d)Load voltage with proposed FVPD controller under phase unbalance

Table 5.6: Phase Unbalance compensation with different controllers

Controller	Phase differences					
	Before Compensation (*between A & B phases*)	After Compensation (*between A & B phases*)	Percentage Error	Before Compensation (*between A & C phases*)	After Compensation (*between A & C phases*)	Percentage Error
FVPD	100^0	120^0	0.00 %	220^0	240^0	0.00 %
IRPT	100^0	107^0	10.83 %	220^0	226^0	5.83 %
SRF	100^0	106^0	11.66 %	220^0	228^0	5.00 %

Table 5.6 represents the quantitative analysis of different controllers under unbalanced phase conditions. With the proposed FVPD controller, phase differences can be restored back to 120^0 and 240^0, as seen above.

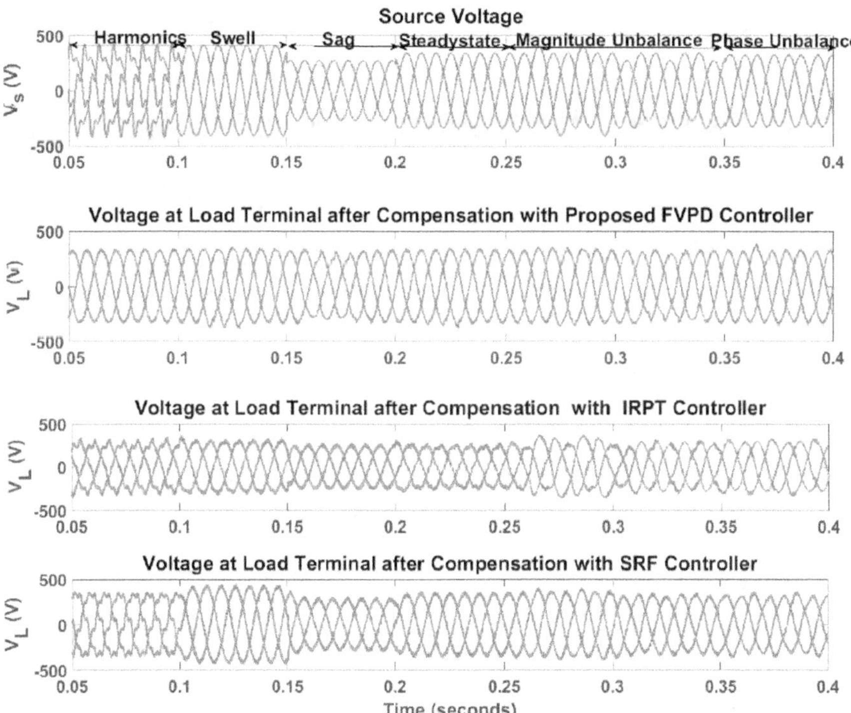

Fig.5. 58. Source voltage and voltage at load terminals after compensation with proposed FVPD, IRPT and SRF controller

Fig.5.58 represents the source voltage and voltage at the load terminal after compensation with proposed FVPD controller, IRPT and SRF controllers respectively. Different voltage abnormalities like harmonics, three phase swell, sag, magnitude unbalance and phase unbalance are marked as occurring in the system voltage profile at different time intervals in the same plot (Fig.5.58). In all these cases, the voltage at the load terminals is maintained at rated value and sinusoidal in shape with the corrective action of proposed FVPD controller whereas with other two controllers, compensation was not properly achieved.

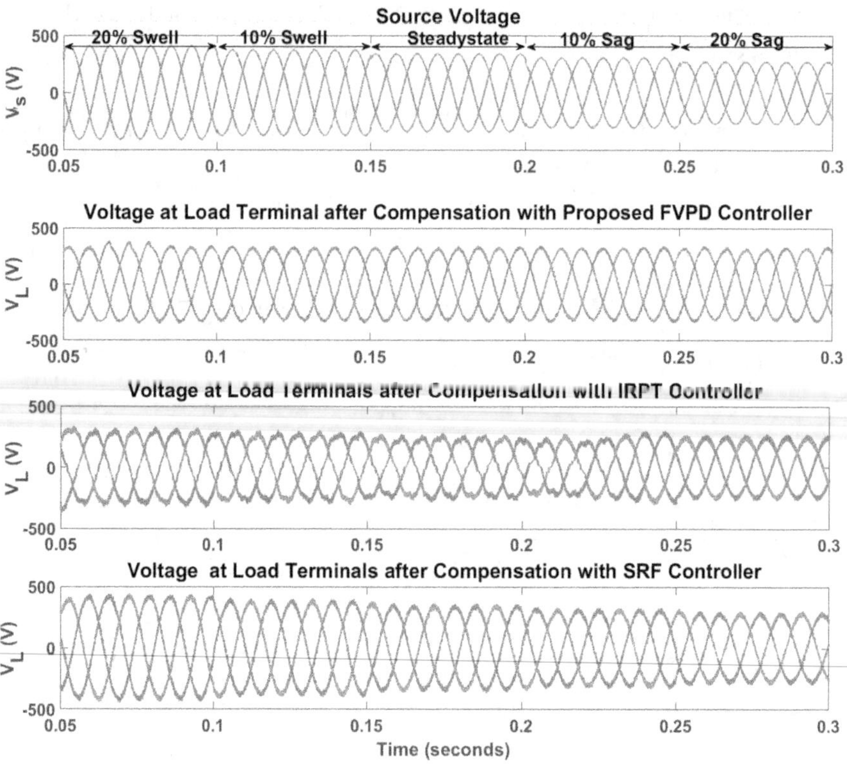

Fig.5.59. Source voltage and voltage at load terminals after compensation with proposed FVPD, IRPT and SRF controller under different sag and swell conditions

In Fig.5.59 the performance of different controllers like proposed FVPD, IRPT and SRF under varying swell and sag conditions are illustrated. The variation considered was ±20% in steps of 10%. Performance of the controllers are checked for 10% of sag and swell, 20% of sag and swell and normal steady state condition (rated voltage condition). It is found that in each of these test condition, the proposed FVPD controller was working effectively compared to other two controllers.

Performance comparison of proposed FVPD controller with already existing controllers like IRPT and SRF has been done and the results are summarized and shown in Table 5.7.

Table 5.7: Simulation Result Analysis

Performance Characteristics	Proposed FVPD	IRPT	SRF
THD	3%	10.43%	12.9%
Voltage Sag	Satisfactory	Not satisfactory	Not Satisfactory
Voltage Swell	Satisfactory	Not satisfactory	Not Satisfactory
Amplitude Unbalance	Satisfactory	Not satisfactory	Not Satisfactory
Phase Unbalance	Satisfactory	Not satisfactory	Not Satisfactory
Number of Computations	Less	More	More
Circuit simplicity	simple	Complex	Complex

Table 5.8 compares hardware components required for implementation of the IRPT, SRF and FVPD controller options.

Table 5.8: Result analysis in terms of hardware components

Hardware Components	Proposed FVPD	IRPT	SRF
Number of multipliers	6	12	10
Voltage sensor of power circuit	6	6	6
Current sensor of power circuit	-	3	-

Number of components required for hardware implementation of proposed FVPD controller is noticeably less which renders it cost-effective.

The combined results of simulation and hardware proves that the proposed FVPD controller effectively compensates all the voltage disturbances and stabilizes the voltage at load terminal with almost instantaneous response time. Moreover, the implementation of the proposed controller is very simple and cheap.

5.8 Conclusion

A novel FVPD controller is explained and its performance is compared with already existing controllers like IRPT and SRF control algorithms. Performance of the proposed controller was evaluated by simulation as well as testing and analysis on a hardware prototype. Both simulation and experimental results validate the effectiveness of the controller, to seamlessly handle transients in source voltage, sag, swell, harmonics, magnitude and phase imbalance, with almost instantaneous response.

CHAPTER 6

DESIGN AND DEVELOPMENT OF CUSTOM POWER PARK

6.1. Introduction

The total number of cumulative hours of interruption was the major benchmark of reliability over a decade ago. Later, even a voltage sag of 20% or less and short durations of interruption were considered to be new measures of reliability. The use of sensitive loads like programmable logic controllers, micro controllers, adjustable speed drives, robotics, loads in hospitals and semiconductor industries are greatly affected by even a minute voltage disturbance. The concept of custom power park eliminates the power quality issues created by the integration of renewable energy sources in micro grid and penetration of large nonlinear devices in a much more efficient and economical way than the customer's own solutions. Basically, the CPP is intended to protect the consumers from power quality issues originating from utility or arising from loads and getting transferred to other customers via utility. This chapter contains the development and testing of various custom power devices employed in the park to improve the power quality.

6.2. Custom Power Park Model

As mentioned in section 4.2 of chapter 4, CPP addresses the voltage disturbance and interruption and other PQ issues like sags, swells, harmonics, notches, unbalance, transients etc. The consumers can get high quality power at a higher cost, which many industrial and commercial customers will be willing to pay, if the expensive production losses could be avoided. Basically custom power park involves the integration of various custom power devices in distribution system so as to provide a guaranteed improvement in the quality of electrical service to the tenants of the park based on their request. It has significant benefits in quality and reliability compared to standard electrical service provided by the utility. On a long term basis, the better choice is the implementation of custom power park due to the reasons like better serviceability, reliability constraints, and most of the voltage sags and interruptions results from events that involves the utility which need to be rectified etc.

The major benefits of custom power park involve the following:

1. No need of installation of power factor correction capacitors by the consumers which may create harmonic resonance and over voltage protection. Power factor correction would be the responsibility of CPP

2. Continuous monitoring of power system conditions will happen. If customers load characteristics change the consumer may be moved to different custom power rate classification.

3. The customers no longer need technical experts or consultants on their payroll. The CPP will provide that.

4. Consumers can select the desired level of service based on their need so as to avoid costly downtime as well as savings achieved by not purchasing, operating, and maintaining internal power conditioning equipment.

5. If customers install individual power conditioning equipment, it involves high cost per power unit, high power losses, need of big space etc. All this can be readily overcome by CPP.

Any cost comparison of Custom Power service with standard service by the utility must include both investment cost and operating cost of "doing it alone" versus the premium rate charges. The operating cost involves cost of losses, maintenance costs, communication costs, spare parts inventory and indirect cost associated with the equipment which includes air conditioning, lighting, floor space, fire protection/monitoring etc. The cost of providing custom power may include a loss of business insurance premium to cover utility liability exposure. Various financial factors need to be considered in the deployment of CPP. The basic idea involves buying power from various power generation companies and improve the power quality and provide different categories of power in terms of improved power quality to the end users based on their choice of category. The cost per unit charged by the power generation companies will be different based on the demand. Also the power quality improvement process involves cost. The resulting output power obviously will be costlier than grid power but better in terms of power quality. As the research work involves the development of CPP, it is largely assumed that the work will also cover fixing up the parameters to be considered for selecting the right power generation companies to optimize the financial burden. Similarly, to maximize the profit, the unit per cost of different grades of power need to be fixed efficiently. But in the Indian context, the deployment generally involves government or government supported agencies and the research work cannot influence largely the layout or fixing tariffs or selection of power generation companies. Hence those aspects are not being covered in the work. The focus of the work thus primarily moves around the power quality improvement techniques and the overall controller for CPP, basically the design, operation and control of a CPP in a most effective way considering all aspects of supply variations/disturbances as well as varying load conditions.

The single line diagram of a custom power plant is shown in Fig.6.1. Here, the power to the custom power park is provided through two feeders from two different substations. The feeders are known as preferred feeder and alternate feeder, which is usually the back up. These preferred and alternate feeders are connected to CPP bus via a solid-state transfer switch (SSTS). This SSTS switches over from one feeder to other and vice versa in a make before break or break before make fashion if fault occurs on any one of the feeders thereby maintaining the continuity of the supply to the loads in CPP load network. It also protects the loads from sags and swells that appear in the utility voltage due to faults. SSTS employs two antiparallel connected thyristor switches. Usually the transfer time is less than a cycle. The performance of SSTS can be improved by replacing thyristor switches with GTOs.

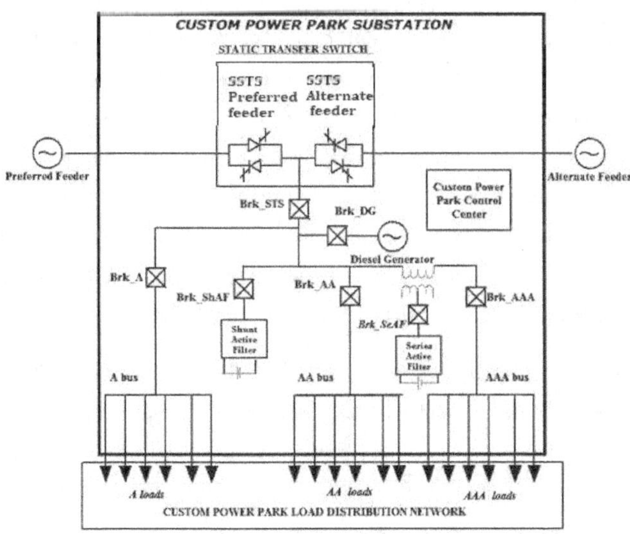

Fig.6.1. Single line diagram of CPP model [26]

The SSTS always ensures that the feeder with higher voltage is selected in less than a cycle in the case of a voltage dip. The two compensating devices used in custom power park are Shunt Active Filter/STATCOM and Series Active Filter/DVR. Shunt Active Filter serves as protection of the system from power pollution caused by nonlinear loads. Shunt Active Filter is basically a controllable current source which when operated in current control mode eliminates all the current related issues like harmonics, unbalance, low power factor etc. The Shunt Active Filter is nothing but a shunt connected voltage source inverter which injects currents into AC system. The difference between the VSI output and system AC voltage causes the current flow and corresponding reactive power exchange. The DC input to the VSI is provided by DC link capacitor, which is converted into AC by proper switching of the power electronic interface used. The injected current tracks the reference currents generated by the control algorithm from sensed load currents and system voltages. Real power exchange is possible if DC link is connected with battery support or RES support which serves as backup power supply. When operated in voltage control mode Shunt Active Filters can provide voltage support in the case of voltage sag. Another power electronic interface used in the park is the Series Active Filter/DVR which injects three phase voltages in series with the line through coupling transformer to make the load side voltage free from voltage distortion, dips, overvoltage and unbalance by exchanging real and reactive power with the line. A rechargeable energy storage system can be connected to the DC terminals to provide additional power support to ride through deep sags and also make real power exchange possible. The inverter output voltage tracks the reference voltage generated by the control algorithm. By injecting the voltages of controllable magnitude, frequency and phase angle, the Series Active Filter can restore the quality of voltage at load terminals when the quality of the source side terminal voltage is significantly out of specification for sensitive load equipment. The combination of Shunt and Series Active Filter known as UPQC which eliminates all current and voltage related issues ensuring protection of all loads in the park, can also be employed.

The power supplied to the consumers by the CPP is categorized into three different grades. They are Grade A, Grade AA and Grade AAA categories. Even though this classification is purely based on the quality of power, these different grades are also different in terms of the cost. The cost per unit will be different for different grades of power so as to meet the expenses incurred by the CPP service provider during power quality improvement on the grid power and also to maximize the profit. These rates will be higher than the regular unit cost as the scheme involves buying of power from various power generation companies which sell power at different rates. Also the processing cost or the operational cost of the CPP will add up to the total money spent. These factors influence the unit per cost of the various grades of power provided by the CPP. Grade A category itself is improved power compared to the standard power delivered by the utility. It gets the services of SSTS and Shunt Active Filter. SSTS protects the loads from system side faults and Shunt Active filter solves all current related issues. So basically, Grade A category loads get harmonic free power with unity power factor at source side. Grade AA category is superior to Grade A category as it gets the services of DG in addition to the services from SSTS and Shunt Active Filter. When both the feeders fail, then the DG can be brought into service after 20 seconds needed for starting. Grade AAA category receives the premium power. This includes all the features of Grade AA. In addition to this, it has a benefit of receiving distortion free and variation free voltages at load terminals with the services offered by the Series Active Filter.

Under normal operating condition the DG stays off and both Grade A and Grade AA receives same improved quality power. Due to some emergency if both feeders fail, then supply to customers of both these categories will be removed through circuit breakers. However, once the DG is ON, the power to consumers of Grade AA category resumes. Grade AA category loads do not lose power more than 20 seconds which is the time taken by the DG for synchronization. For Grade AAA category, continuity of the power will not be lost even during those 20 seconds because of Series Active Filter/DVR action. The power to Grade A category will be restored only when any one of the feeders is back into service. Since the Series Active Filter/DVR has to provide voltage support for 20seconds during DG start up, the DC link capacitor should be capable of holding the charge for such a long duration. This can be ensured by connecting a battery backup or renewable energy supported battery backup at DC link terminals. The grade of the quality of power a customer in the park receives depends on the nature of its load and the price he is ready to pay.

The centralized overall controller coordinates different devices in the CPP in a flexible way. The ON-OFF status of various devices shown in Fig.6.1, for different conditions are shown in Table.6.1. The preferred and alternate feeder voltages are monitored and power quality events are captured and managed for periodic assessment of the service being provided.

Table.6.1: ON-OFF states of CPP devices and loads under different conditions

Case		SSTS-preferred feeder	SSTS-alternate feeder	Diesel Generator	Series Active Filter	Shunt Active Filter	A Loads	AA Loads	AAA Loads
<10% sag at preferred and alternate feeder	i)Vp>Va	On	Off	Off	On	On	On	On	On
	ii)Vp<Va	Off	On	Off	On	On	On	On	On
Between 11%-40%sag/swell at both feeders	i)Vp>Va	On	Off	Off	On	On	On	On	On
	ii)Vp<Va	Off	On	Off	On	On	On	On	On
<10% sag at preferred and Between 11%-40% sag/swell at alternate feeder		On	Off	Off	On	On	On	On	On
<10% sag at alternate feeder and Between 11%-40%sag/swell at preferred feeder		Off	On	Off	On	On	On	On	On
<10% sag/swell at preferred feeder&>40% sag at alternate feeder		On	Off	Off	On	On	On	On	On
>40% sag/swell at preferred feeder&<10% sag at alternate feeder		Off	On	Off	On	On	On	On	On
>40% sag or interruption at preferred &alternate feeders during synchronization delay		On	Off	Off	On	Off	Off	Off	On
>40% sag or interruption at preferred &alternate feeders after synchronization delay		Off	Off	On	On	Off/On	Off	On	On

After monitoring the voltages when both preferred and alternate feeder voltages are between 0-40% sag, and whichever is at higher voltage, that feeder will be switched on. If preferred feeder voltage is less than 10% and alternate feeder voltage is between 11-60% then alternate feeder will feed the loads and vice-versa. In all these cases, all the three categories of loads, ShAF and SeAF will also be on. DG will be off. These scenarios occur during distribution side faults. If the voltage at both the feeders are less than 60%(greater than 40% sag), as in the case of transmission line faults, then DG will be on. During the time taken for synchronization, the loads A and AA will not receive any power. Shunt Active Filter also will be off. SeAF provides power to load AAA during this time. After synchronization delay, the power to load AA also will be restored. Usually this being considered as an emergency situation, ShAF will be off and DG alone supports the grade AA and grade AAA categories without considering the quality improvement of the voltage profile. However, with good support from the optional battery backup connected at the DC link of ShAF (the condition arising when renewable energy is available in abundance) then the ShAF can also acting as a power source supporting DG to power the AA and AAA category CPP loads. In addition to power support, ShAF when in the circuit also provides current harmonic elimination and thereby power quality improvement. In this way, all the CPDs and loads coordinate to provide improved quality and reliable power.

6.3. Coordinated Controller for Custom Power Park

The CPP contains different types of network reconfigurable and compensating devices. A proper coordination of these devices and various loads in the park is essential for proper working of CPP to deliver improved PQ and reliable power to consumers. This coordination among different devices and loads can be smoothly done using an overall coordinated controller in the custom center.

The voltages at the two feeders are sensed and given to the overall controller which decides and sends signals to the breakers connected to different loads, DG, Shunt and Series Active Filters. Overall controller also sends signals for proper switching of SSTS. Fig.6.2. represents the flowchart for the overall coordinated controller. The voltages at preferred (Vp) and alternate (Va) feeders are sensed and if it is greater than 0.6pu then again one more condition is checked to determine which feeder is having higher voltage. If Vp is higher, then the status of flags for preferred (cp) and alternate feeders (ca) will be checked. If cp=0 and ca=0, then both feeders are not ON. Switching of preferred feeder needs to be done and all loads, ShAF, SeAF need to be ON. If cp=1 and ca=0, then preferred feeder is already ON. If cp=0 and ca=1, which indicate alternate feeder is ON, then switchover from alternate feeder to preferred feeder is required and corresponding loads and CPDs also need to be ON. If Va is higher then also the flag status will be checked and necessary, task will be performed as explained previously. If the voltage at preferred and alternate feeders is less than 0.6, then DG status flag will be checked. Zero indication of this flag means DG is not on and the DG start signal will be sent and response awaited. Once DG is ready, the loads AA and CPDs will be ON and DG starts powering AAA loads also rather than using SeAF to provide real power. During the time needed for the generator to come into action, Series Active Filter supplies power to AAA load.

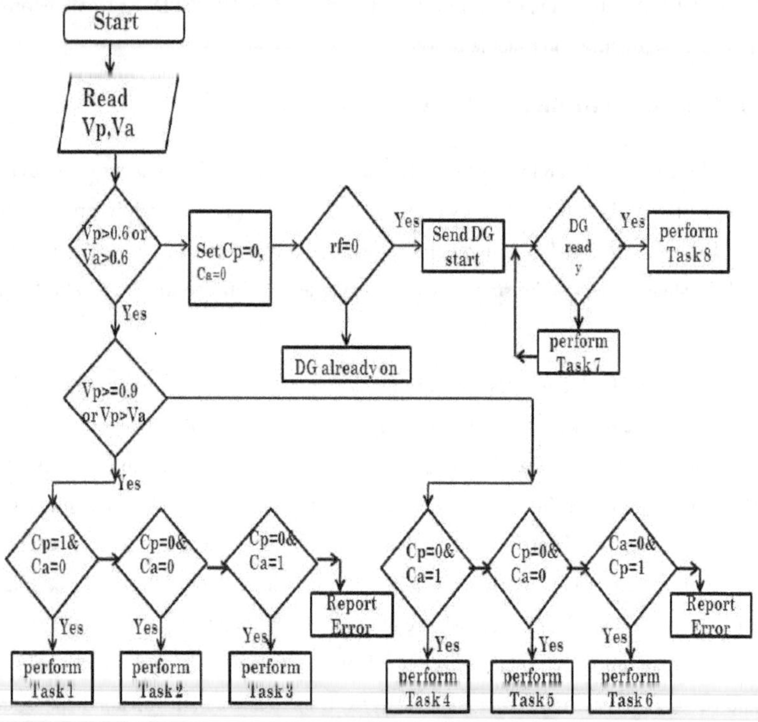

Fig.6.2 Flow chart illustrating the operation of coordinated controller

The algorithm corresponding to Fig.6.2 can be written as

i. The voltages at preferred (Vp) and alternate (Va) are sensed

ii. If Vp and Va are greater than 0.6 go to step 3

iii. Check Vp>Va, if yes, go to step 4

iv. If the status of flags for preferred (cp) = 0 and alternate feeders (ca) = 0, then it means that both feeders are not switched ON. Switching ON of preferred feeder needs to be done and all loads, ShAF, SeAF need to be switched ON

v. If cp = 1 and ca = 0, then preferred feeder is already ON

vi. If cp = 0 and ca = 1, this indicate alternate feeder is ON, then switchover from alternate feeder to preferred feeder is required, and corresponding loads and CPDs also need to be switched ON

vii. If Va is higher than Vp then also the flag status will be checked and based on the status of the flags corresponding task (as mentioned in Table 6.2) will be performed.

viii. If the voltage at preferred and alternate feeder is less than 0.6 pu, then DG status flag will be checked. Zero indication of this flag means DG is need to be switched ON and the DG start signal will be sent from Coordinated Controller to DG and response awaited. Once DG is ready, (signal send to indicate DG is ready) the AA, AAA loads, and Custom Power Devices(CPDs) will be on. During the time needed for the generator to come into action, Series Active Filter supplies power to AAA load.

The Table 6.2. shows the status of circuit breakers connected to loads and different devices in the power park for different cases considered in the flowchart for proper switching of CPDs and loads by the control signal generated by overall controller. For each case what needs to be displayed is also clearly mentioned in the Table 6.2. Basically the flow chart along with tabular data illustrates the effectiveness of the overall controller in coordinating the devices in the park to ensure proper operation.

Table.6.2: Status of Switches

	Display	Status of Switches							
		SSTS-preferred feeder	SSTS-alternate feeder	Breaker-A Load	Breaker-AA Load	Breaker-AAA Load	Breaker-ShAF	Breaker-SeAF	Breaker-DG
Task 1	Preferred feeder is already on	On	off	On	on	on	on	on	Off
Task 2	Switch on preferred feeder	On	off	On	on	On	on	on	Off
Task 3	Switch over from alternate feeder to preferred feeder	On	off	On	on	On	on	on	Off
Task 4	Alternate feeder is already on	Off	on	On	on	On	on	on	Off
Task 5	Switch on alternate feeder	Off	on	On	on	On	on	on	Off
Task 6	Switch over from preferred feeder to alternate feeder	Off	on	On	on	On	on	on	Off
Task 7	DG not ready	Off	off	Off	off	On	off	on	Off
Task 8	DG is on	Off	off	Off	on	On	Off/On	on	On

6.4. Simulation and testing of proposed Custom Power Park

The Custom Power Park model has been developed in MATLAB/Simulink and the performance of different custom power devices has been analyzed. Fig.6.3. represents the simulation circuit. Two x three phase programmable voltage sources of rating 415V, 50Hz have been considered as the preferred and alternate feeders coming from two different substations. The voltage sources are connected to load network through SSTS which is nothing but two back to back connected thyristor pairs which are controlled by the signals from CPP control centre. Mainly three custom power devices have been considered which are Shunt Active Filter, Series Active Filter and SSTS. The loads are divided into three main categories such as Grade A, Grade AA and Grade AAA. The performance of the custom power devices and overall coordinated controller in providing valuable high quality services to the consumers has been analyzed by carrying out the simulation.

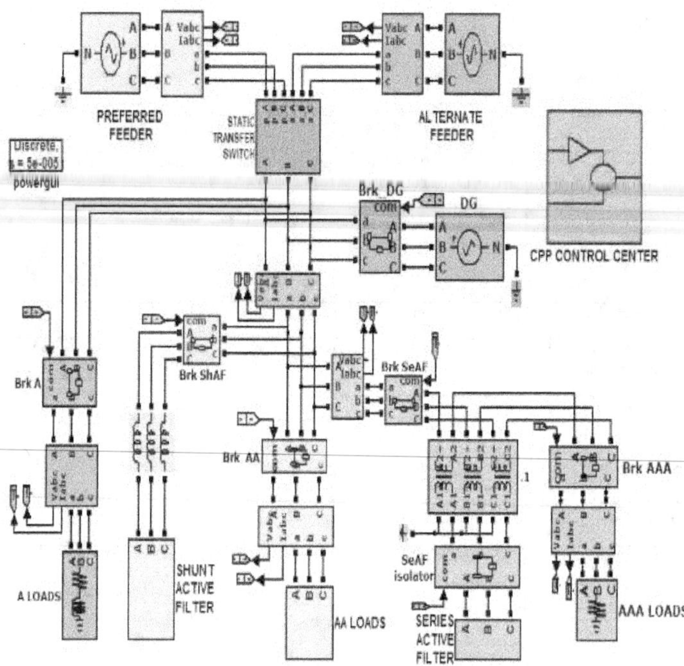

Fig.6.3. Matlab Simulink Circuit of CPP

The load details are given in Table 6.3

Table 6.3: CPP Load details

Feeder Voltage	415V
Grade A Category	10kW linear load(three phase RL Load)
Grade AA Category	10kW,1 kVAR ,Linear load(three phase RL load)+Diode bridge rectifier with R load (nonlinear load)
Grade AAA Category	10kW linear load(three phase RL Load)

Simulation study has been carried out in Matlab Simulink Platform to analyze the working of the custom power devices to improve the quality of power.

6.4.1. Performance Analysis of Shunt Active Filter in Custom Power Park

Shunt Active Filter is mainly deployed to safeguard the loads in the custom power park from disturbances caused by other nonlinear loads in the system. It acts as a shield for the loads in the distribution network of CPP from issues like current harmonics, reactive power consumption, low power factor, unbalance in load currents etc. Shunt Active Filter is a VSI operated in current control mode to achieve the necessary harmonic and reactive power compensation. Various control algorithms have been proposed to generate the reference currents which includes IRPT [56], SRF [59], DC bus voltage control algorithm [58], SD [57] etc. In this simulation study, the control algorithm used is IcosΦ control algorithm [60] which has the advantages like less response time, simple, better performance even under unbalanced source and load conditions, efficient performance under transient conditions etc. The block diagram representation of IcosΦ control algorithm is shown in Fig.6.4.

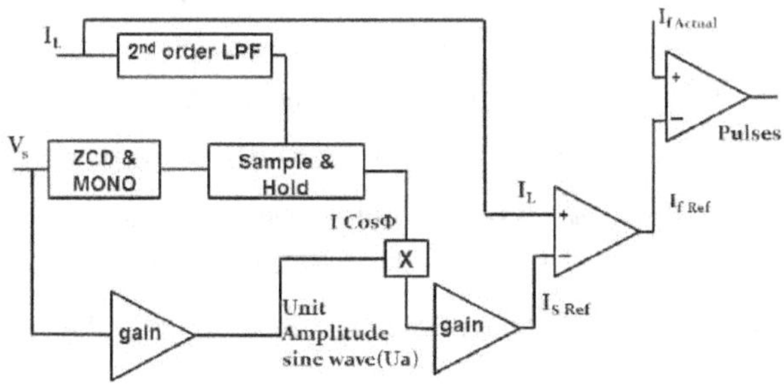

Fig.6.4. Block diagram representation of IcosΦ control algorithm

According to this control algorithm, the source has to supply only the real part of the fundamental load current. The rest (reactive and harmonic components) will be supplied by the filter. The real part of the fundamental load current is nothing but the I cosΦ component and hence the name, IcosΦ control algorithm. Assuming a balanced source, the three-phase instantaneous voltage can be specified as:

$$v_a = V_m.\sin \omega t; v_b = V_m.\sin(\omega t - 120°); v_c = V_m.\sin(\omega t + 120°); \qquad (6.1)$$

The reference compensation currents for the Shunt Active Filter are deduced as the difference between the actual load current and the desired source current in each phase.i.e.

$$i_{Fa\,(Ref)} = i_{La} - i_{sa(ref)}; i_{Fb\,(Ref)} = i_{Lb} - i_{sb(ref)}; i_{Fc\,(Ref)} = i_{Lc} - i_{sc(ref)} \qquad (6.2)$$

Where the desired (reference) source currents in the three phases are given as,

$$I_{sa(ref)} = |I_{s(ref)}| \times U_a = |I_{s(ref)}|.\sin \omega t; \qquad (6.3)$$

$$I_{sb(ref)} = |I_{s(ref)}| \times U_b = |I_{s(ref)}|.\sin(\omega t - 120°); \qquad (6.4)$$

$$I_{sc(ref)} = |I_{s(ref)}| \times U_c = |I_{s(ref)}|.\sin(\omega t + 120°); \qquad (6.5)$$

U_a, U_b and U_c are the unit amplitude templates of the phase to ground source voltages in the three phases respectively.

$$U_a = 1.\sin \omega t \; ; \quad U_b = 1.\sin(\omega t - 120°); U_c = 1.\sin(\omega t + 120°); \qquad (6.6)$$

The magnitude of the desired source current $|I_{s(ref)}|$ can be expressed as the magnitude of the real component of the fundamental load current in the respective phases (I cos Φ), .i.e. for 'A' phase it can be written as $|I_{s(ref)}| = |\text{Re}\,(I_{La})|$. In the case of unbalance, the average of the magnitude of real components will be taken before multiplying with unit vector templates.

A zero crossing detector (ZCD) is used to detect the negative going zero crossing of the corresponding phase voltage (V_S). The ZCD has been designed with a tolerance of 5% to ensure that any oscillations around the zero-crossing are taken care of. The phase-shifted fundamental current goes as the "sample" input and the ZCD output pulse goes as the "hold" input to the "sample and hold" circuit whose output is the magnitude $|I_{s(ref)}|$. This magnitude is multiplied with unit amplitude sine wave (U_a) to get the reference source current (I_{SRef}). The reference source current when subtracted from load current (I_L), gives the reference filter current ($I_{f\,Ref}$). The reference filter current and actual filter current ($I_{f\,Actual}$) are given as inputs to the hysteresis controller to get the pulses for the inverter.

Fig.6.5 represents waveforms depicting the performance of Shunt Active Filter.

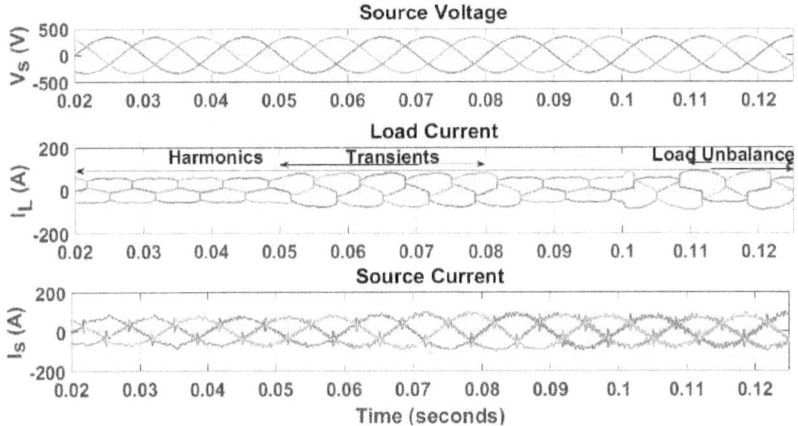

Fig.6.5. Source Voltage, Load Current and Source Current waveforms

From 0.05 to 0.08 sec load transients are present in the system. During the period 0.1 to 0.15 secs, a single phase load is switched on which creates the unbalance in the load current and a diode bridge rectifier is turned on from starting onwards which draws highly nonlinear current. In all these conditions, the ShAF gives satisfactory performance maintaining sinusoidal source current at unity power factor. The increase in the source current amplitude is due to switching on of the extra loads into the system. The Shunt Active Filter with IcosΦ controller efficiently eliminates all the current related problems like harmonics, unbalance, transients and low power factor etc when operated in current control mode in less than half a cycle. All the loads in the CPP distribution network gets services of Shunt Active Filter. So with IcosΦ controller the currents in the CPP bus will be sinusoidal in shape and in unity power factor with the bus voltage.

6.4.1.1. Transients:

When load transients occur the Shunt Active Filter with IcosΦ controller, immediately comes into action so that there will be a smooth transition from one state to another.

Fig.6.6 represents the waveforms of source voltage, load current, reference filter current and source current before and during the load transient condition. At 0.05 sec, load is increased and the corresponding increase can be seen in reference filter current and source current also. The response time of ShAF with IcosΦ controller during dynamic condition is found to be very less which is of 0.002seconds.

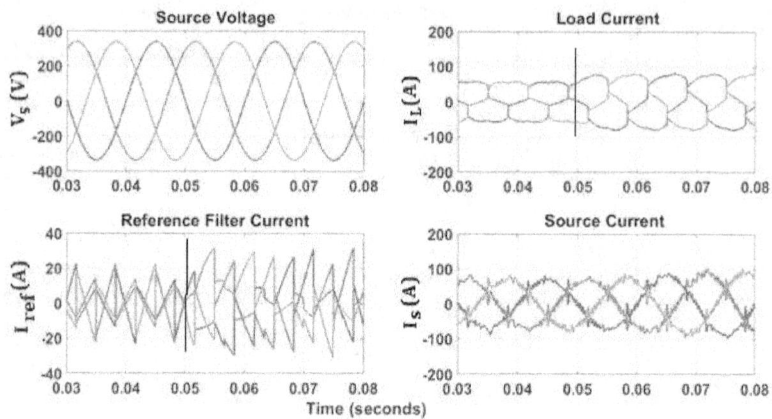

Fig.6.6. Waveforms of source voltage, load current, reference filter current and source current under transient condition.

6.4.1.2. Current harmonics:

The penetration of RES and EV charging units into the grid makes the current distorted. In addition to this the nonlinear loads composed of power electronic devices also draw distorted current from grid. The ShAF with IcosΦ controller eliminates the current harmonics and maintains unity power factor at source side. Fig.6.7. shows the waveforms of source voltage, load current, reference filter current and source current under harmonic conditions. It is observed from the Fig.6.7 that the current harmonics are eliminated by the action of ShAF thereby pure sinusoidal waveform is obtained at source end.

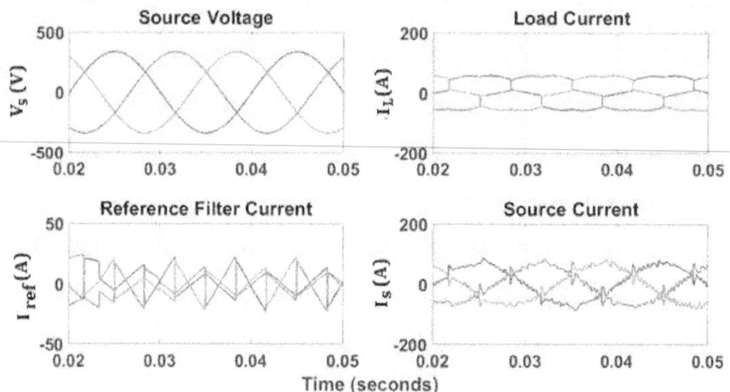

Fig.6.7. Waveforms of source voltage, load current, reference filter current and source current under harmonic condition.

Fig.6.8 depicts the source voltage and source current waveforms before and after compensation. Before compensation the source current is similar to load current which is distorted and not in phase with the source voltage. The ShAF effectively compensates the harmonics and the source current becomes in phase with the source voltage thereby maintaining unity power factor. The reference and actual filter currents are exactly matching as shown in Fig 6.9 to carry out the necessary compensation so that all the current related issues can be sorted out effectively with the help of IcosΦ controller which is simple and efficient compared to other controllers used for ShAF.

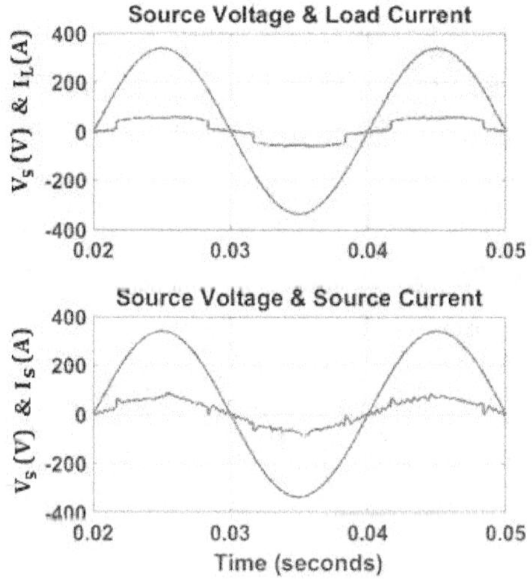

Fig.6.8. Source voltage and source current waveforms before and after compensation.

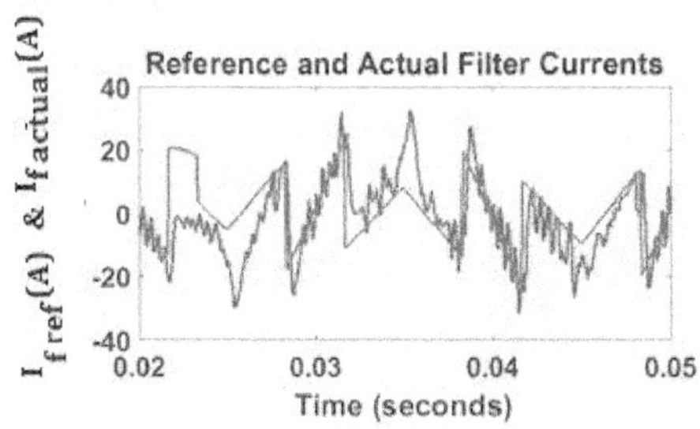

Fig.6.9. Reference and Actual Filter Currents

6.4.1.3. Unbalance:

The performance of the ShAF in this case is very crucial as unbalance can arise in the power system at any time. In the preferred controller the average of the three phase currents has been taken to generate reference current magnitude so that unbalance condition can be taken into account. Fig.6.10 represents the source voltage, load current, reference filter current and source current waveforms in which the currents become unbalanced at 0.1sec. The corresponding changes in reference filter currents are also seen so that the source current become balanced sinusoidal ones even during the unbalance period.

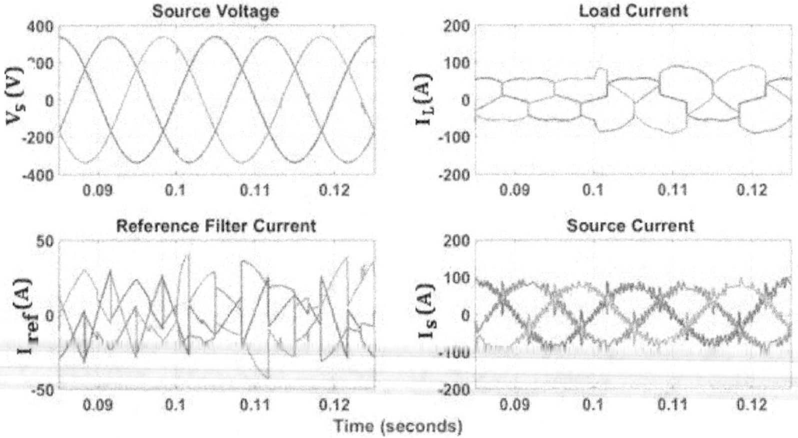

Fig.6.10. Waveforms of source voltage, load current, reference filter current and source current under unbalance condition

From the above results it is concluded that the ShAF with IcosΦ controller performs very effectively under the conditions of transients, unbalance, current harmonics etc to compensate these abnormalities and thereby maintains upf and sinusoidal current at the source side with in a fraction of cycle.

6.4.2. Performance Analysis of Series Active Filter in Custom Power Park

Series Active Filter ensures the safety of sensitive loads in the custom power park. It protects the sensitive loads from voltage perturbations like voltage sag, swell, unbalance, phase unbalance and voltage harmonics and thereby maintains load terminal voltage with stipulated magnitude and sinusoidal shape. The control algorithm used for SeAF for performance analysis is the proposed Fundamental Voltage Peak Detection (FVPD) algorithm which has been already explained in chapter 5.

6.4.2.1. Voltage Harmonics:

The nonlinear current drawn by the nonlinear loads causes distortion in the supply voltage. The SeAF eliminates the voltage harmonics by injecting a voltage in series with the line through a series coupling transformer.

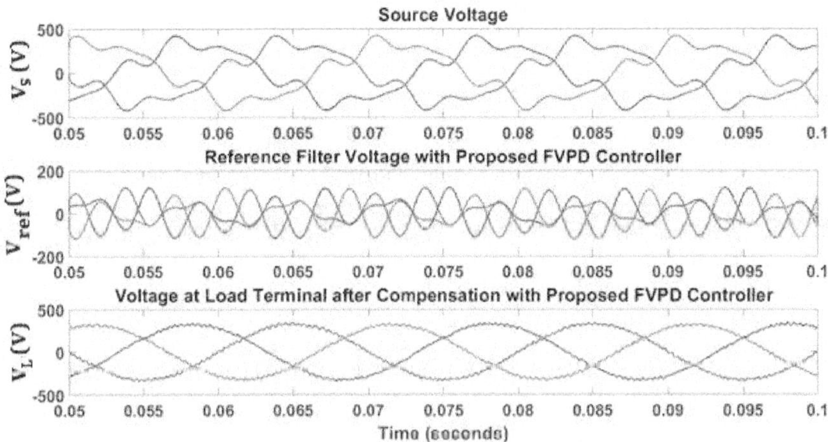

Fig.6.11. Waveforms of source voltage, reference filter voltage and load terminal voltage under voltage harmonic condition

Fig.6.11. shows the source voltage, reference filter voltage and voltage at load terminals respectively. The source voltage contains fifth and seventh order harmonics which are eliminated by the action of SeAF with FVPD controller and at load side a sinusoidal voltage of rated magnitude is obtained. Fig.6.12. represents the reference and actual filter voltages (voltage at the secondary side of the injection transformer) which as exactly same which in turn indicates perfect compensation.

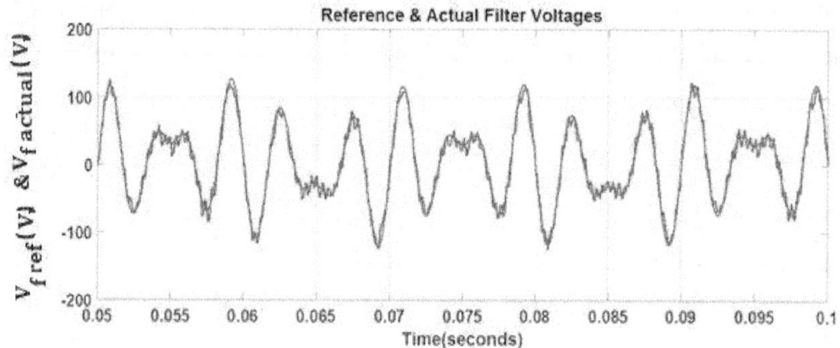

Fig.6.12. Reference Filter and Actual Filter Voltages

Fig.6.13 represents the FFT analysis of the voltage before and after compensation. Before the compensation the THD was 25% which is reduced to 3.5% by the action of SeAF. After the compensation the THD is very well within the limit specified by IEEE 519 standard.

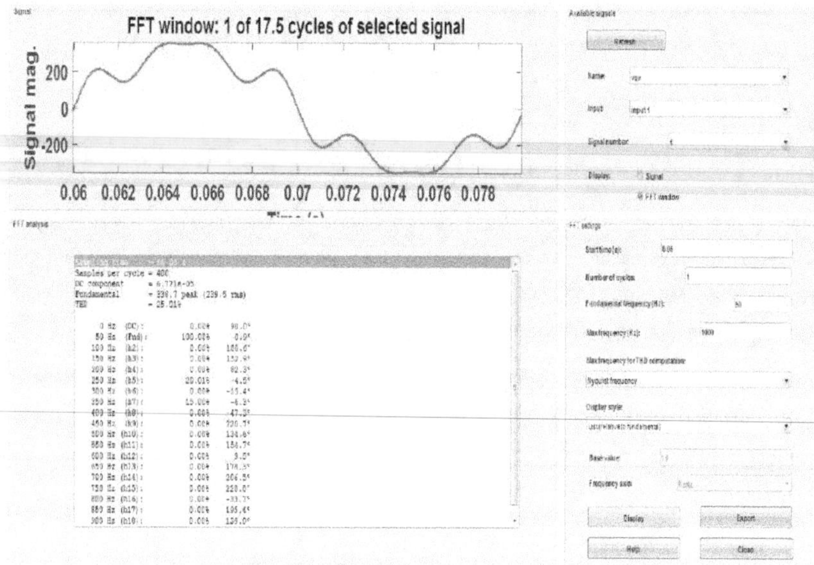

Fig.6.13. (a)FFT analysis before compensation

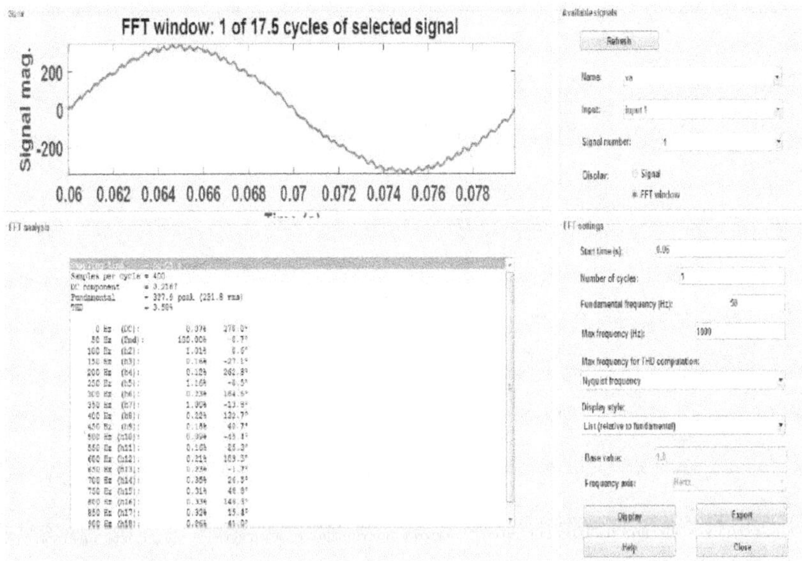

Fig.6.13. (b)FFT analysis after compensation

6.4.2.2. Voltage Sag:

Occurrence of fault in the feeders of custom power park may cause a voltage sag. Fig.6.14 represents the waveforms in the case of three phase sag. The percentage sag considered for simulation analysis is 20%. In the case of sag, the voltage injected by the SeAF is in phase with the system voltage so that both will get added resulting in rated voltage at the load terminal which is clearly seen in Fig.6.14. Till 0.25sec it is rated voltage with corresponding compensating voltage as zero and at t=0.25sec, voltage sag is created. Accordingly compensating voltage is also generated by proper controller action which when added with source voltage brings the load terminal voltage back to rated value.

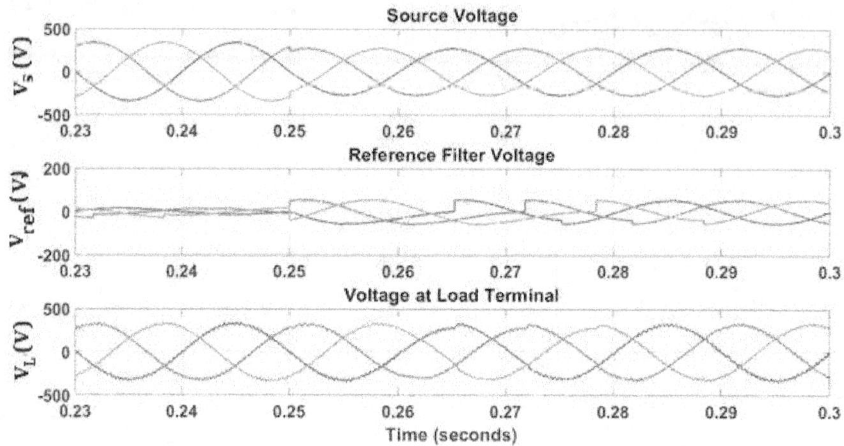

Fig.6.14. Waveforms of source voltage, reference filter voltage and load terminal voltage under voltage sag

6.4.2.3. *Voltage Swell:*

Faults on feeders and sudden removal of loads give rise to the occurrence of voltage swell in the custom power park. SeAF injects voltage which is 180^0 out of phase with the system voltage so that sum of source voltage and injected voltage reduces the voltage at the load terminal and brings it back to rated value which is clearly depicted in Fig.6.15. Swell is introduced at 0.15sec. Till that time the reference voltage is found to be zero.

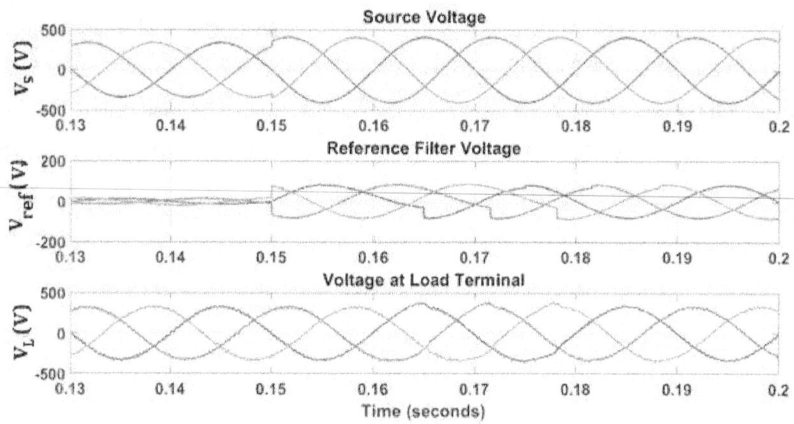

Fig.6.15. Waveforms of source voltage, reference filter voltage and load terminal voltage under voltage swell

6.4.2.4. Amplitude Unbalance:

To analyze the performance of SeAF with FVPD controller under amplitude unbalance, a 20% increase in voltage is created in R phase (blue colour) alone during t=0.35sec to t=0.4sec as shown in Fig.6.16. The controller generates a reference voltage corresponding to that phase to correct the unbalance and thereby prevents the reflection of unbalance in load terminal voltage.

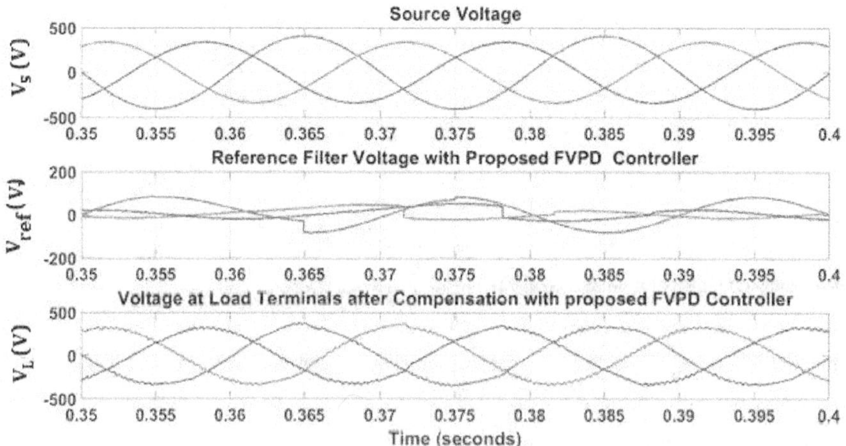

Fig.6.16. Waveforms of source voltage, reference filter voltage and load terminal voltage under amplitude unbalance

6.4.2.5. Phase Unbalance:

To check the performance of the SeAF with proposed FVPD controller under phase unbalance condition, phase difference between A and B is changed to 100^0 and between A and C is changed to 220^0 instead of 120^0 and 240^0 respectively. Compensating voltage is generated by the filter for the two phases as seen in Fig.6.17, which when added with system voltage corrects the phase difference to 120^0 and 240^0 as needed.

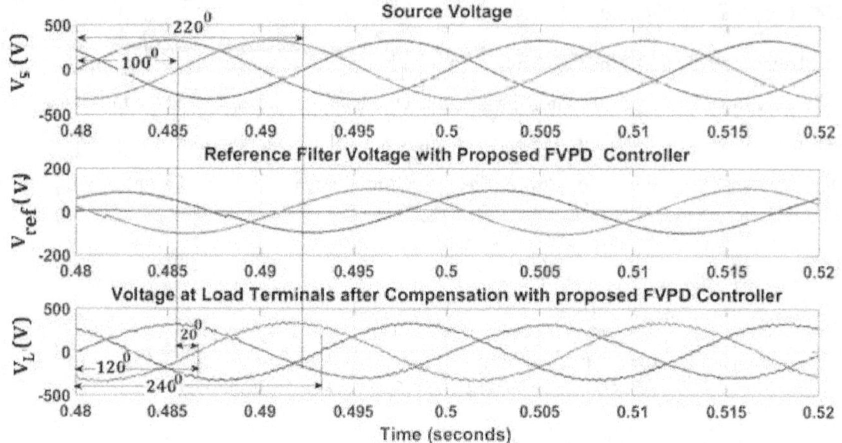

Fig.6.17. Waveforms of source voltage, reference filter voltage and load terminal voltage under phase unbalance

From the simulation analysis of SeAF with proposed FVPD controller, it has been proved that it performs well under perturbations in voltage like sag, swell, harmonics and amplitude and phase unbalance thereby maintaining rated sinusoidal voltage at the load terminals. This ensures the protection of sensitive loads under any kind of voltage disturbance occurring in the park which makes SeAF a crucial element in the custom power park.

6.4.3. Performance Analysis of Custom Power Park with Diesel Generator (DG) /RES Support

Sometimes due to faults in both preferred and alternate feeders, the voltage at both feeders can fall below 60% which is basically low voltage as far as the loads are concerned which are designed to operate at rated voltage. Under this case DG will provide power to AA and AAA category loads. The waveforms associated with the operation of DG is shown in Fig.6.18. It is seen that at 0.2sec, both preferred and alternate feeder voltages become 50%. Immediately the DG starts feeding the loads. There is some small time delay for synchronization of DG. With the help of ShAF and SeAF and with enough battery backup/RES support, the continuity of supply can be maintained during that short time delay too. Mainly DG protects the loads from voltage sag created by faults.

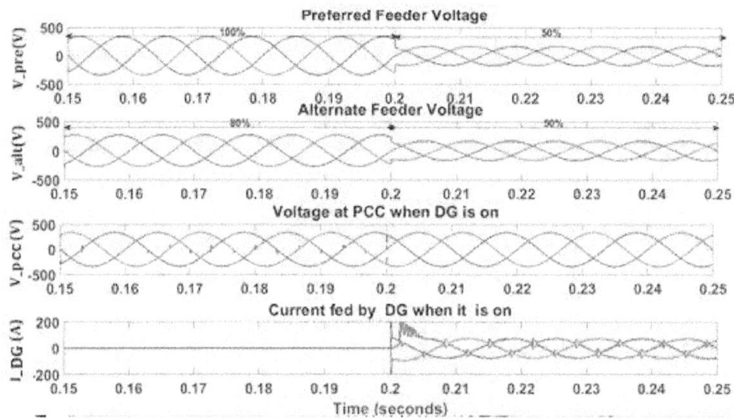

Fig.6.18. Waveforms associated with operation of DG

6.4.4. Performance Analysis of Solid State Transfer Switch (SSTS) in Custom Power Park

SSTS is used to switch over from preferred feeder to alternate feeder and vice versa when the voltage at a particular feeder falls below certain percentage. SSTS comprises of power electronic devices that make it possible to switchover from one feeder to another feeder within 20msec or faster. The static transfer switch solves the feeder voltage issues by switching to that feeder which has better voltage condition closer to rated value, as shown in Fig.6.19. Till 0.1sec, the preferred feeder voltage is 100% and feeds the load. At 0.1sec, preferred feeder voltage is reduced to 80% whereas the alternate feeder voltage is 100%. The same instant, SSTS comes into action by switching over from preferred to alternate feeder and alternate feeder starts feeding the loads. The alternate feeder current waveforms are shown in Fig.6.19. Again at 0.15sec, as the preferred feeder voltage is higher than alternate feeder voltage, switchover to preferred feeder has been done and preferred feeder current waveforms indicates that it started feeding the load. During 0.2 to 0.25 sec the voltages at both feeders have come down to 50% and both the feeders are disconnected from load.

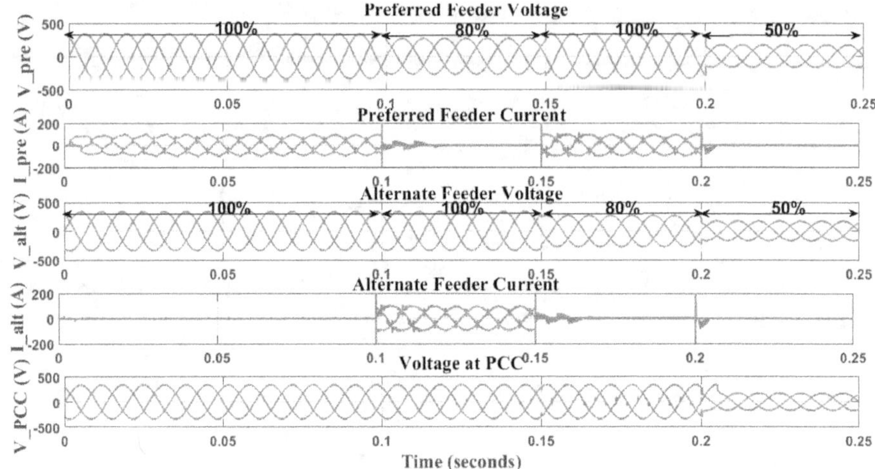

Fig.6.19. Waveforms of preferred feeder voltage and current, alternate feeder voltage and current and voltage at PCC

6.4.5. Performance Analysis of Coordinated Controller in Custom Power Park

The controller for the entire system is modeled using MATLAB coding . The preferred and alternate feeder voltages are sensed and given as input to this controller. The sag detection method used is shown in Fig.6.20.

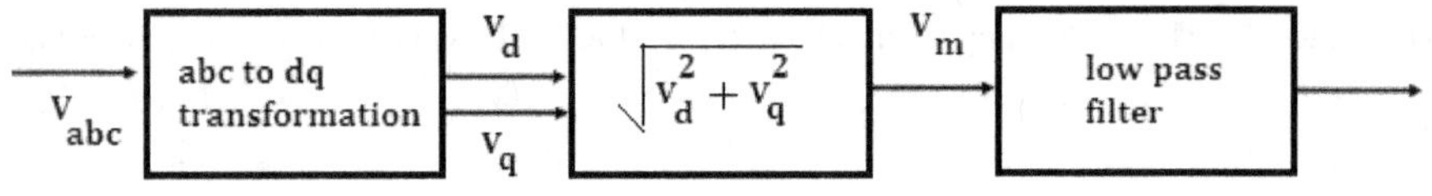

Fig.6.20.Voltage Detection method

The three phase voltages are sensed and converted to dq reference frame. Using a hypotenuse function its magnitude is then calculated and given to a low pass filter to attenuate transients.The output of voltage detection circuit is shown in Fig.6.21. To check the effectiveness of voltage detector, 20% sag and swell is introduced at t=0.1 sec & t=0.2 sec respectively.The detection circuit senses these voltage changes with no delay as seen in Fig.6.21.

The output of voltage detector is given to overall coordinated controller. It checks the voltages on the feeders and switches as per the conditions as shown in flow chart in Fig.6.2. The conditions shown in Table 6.4 are used in the simulation to check the operation and performance of the controller of the park.

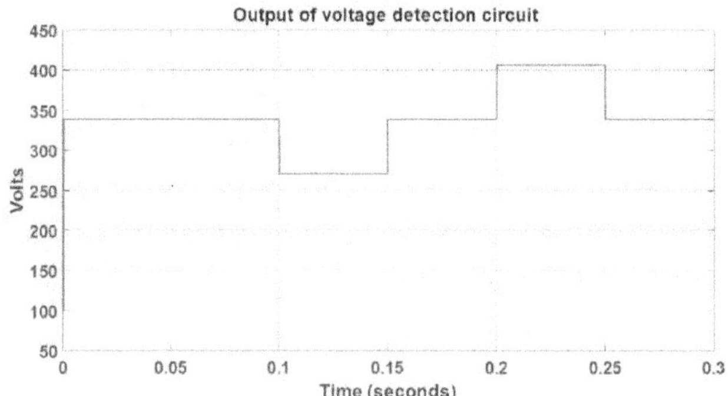

Fig.6.21.Output of voltage detection circuit

Table 6.4: Voltages in preferred and alternate feeders

Time(s)	Preferred feeder voltage	Alternate feeder Voltage
0-0.1	100%	100%
0.1-0.15	80%	100%
0.15-0.2	100%	80%
0.2-0.25	50%	50%

The waveform in Fig.6.22. represents preferred feeder voltage, alternate feeder voltage, and the voltages and currents at Load A (Grade A), Load AA (Grade AA) and Load AAA (Grade AAA) respectively. For the simulation analysis, linear loads are considered for Grade A and Grade AAA category whereas for Grade AA combination of linear and nonlinear loads are considered. The effectiveness of the static transfer switch and controller to prevent the feeder voltage issues by switching to available good voltage feeder or switching the AA and AAA loads (critical loads) to the diesel generator in case of failure of both feeders can be seen in Fig.6.22. Upto 0.1 sec the preferred and alternate voltages are set to nominal values. From $0.1-0.15$ sec, the preferred voltage is reduced to 80% of nominal value which causes switch over of loads to the alternate feeder. At 0.25 sec, both the feeder voltage has dropped to 50%. It is seen that when both preferred and alternate voltages are at 50%, the continuity of supply is lost in Load A which is a grade A type load whereas in Load AA and Load AAA the continuity is maintained using DG. At 0.25 sec, both the feeder voltages are restored and loads are switched to the preferred feeder. At 0.25sec -0.3sec, even though the voltage is distorted at preferred feeder, the voltage at Load AAA (grade AAA) is pure sinusoidal with rated voltage because of the service provided by Series Active Filter whereas the voltages at other two load terminals are distorted. Whatever the circumstances the Grade AAA category loads get pure sinusoidal, rated voltage with the help of CPDs and overall controller. The A loads gets uninterrupted power for all times except the failure of both feeders below 60% of the nominal values. The AA and AAA loads are able to receive uninterruptable power at all the times with the help of SSTS, DG and overall controller of the park.

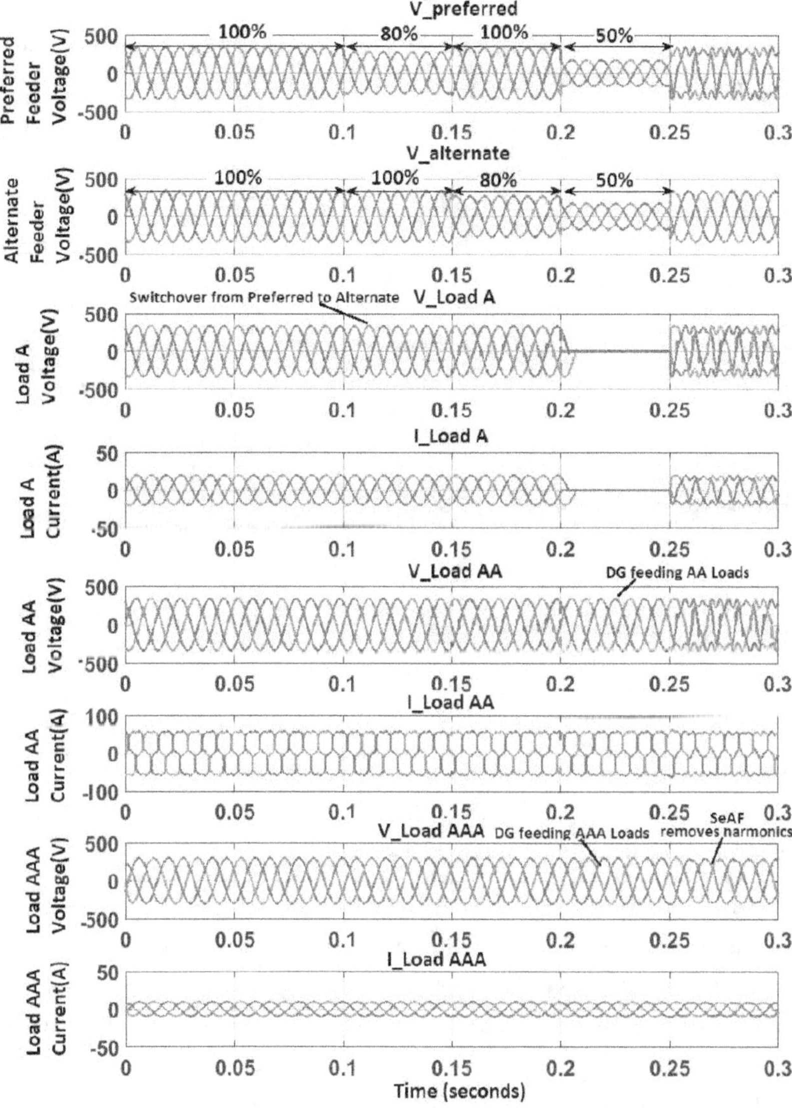

Fig.6.22. Preferred feeder voltage, alternate feeder voltage, voltages and currents at Load A (Grade A), Load AA (with linear load, Grade AA) and Load AAA (Grade AAA)

Fig.6.23 represents the voltage and current waveforms at CPP bus according to the conditions specified in the Table 6.4. Here it is assumed that when voltage at both preferred and alternate feeders are less than 60%, DG is introduced into the bus. Hence rated voltage is seen during 0.2sec to 0.25sec also. The current drawn from bus is also seen sinusoidal with the action of ShAF. Even though the voltage is distorted during the period 0.25s to 0.3s, the current remains undistorted due to proper control action. This clearly depicts that whatever perturbations happen, the other loads connected to the bus will not get affected.

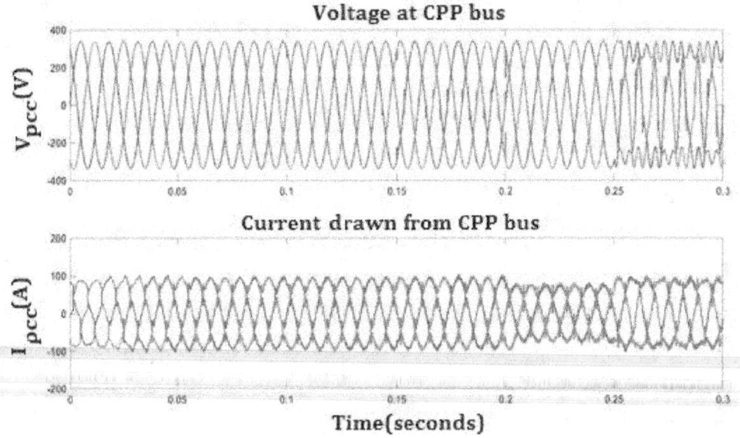

Fig.6.23. Voltage at CPP bus and Current drawn from CPP bus

6.5. Hardware Implementation of proposed CPP

The main parts of the CPP includes SSTS, Shunt Active Filter, Series Active Filter, DG, Overall controller and loads in CPP distribution network. The overall controller coordinates different custom power devices and loads. It produces necessary control signals based on the sensed voltages of the preferred and alternate feeders so as to control and coordinate different devices in the park. The hardware set up mainly consist of sensing of feeder voltages, development of SSTS, development of novel FVPD controller circuit, development of overall controller circuit and communication module development. The overall set up is shown in Fig.6.24. For testing purposes, 60W LED lamps are used as A, AA and AAA category loads.

Fig.6.24. Overall Hardware Setup

6.5.1. Voltage Detection Scheme

The first and foremost step in the testing is signal conditioning. The feeder voltages are connected to step down transformer and then to DF10 series single phase rectifier IC. The output of the rectifier IC is connected to a simple DC voltage sensor which is a voltage divider circuit. The voltage sensor is integrated with Arduino Uno board which reads the sensor output and sends it to the master controller.

6.5.2. Development and testing of SSTS

6.5.2.1. Development of SSTS:

SSTS can be made up of thyristors or GTOs. The choice of particular switch as SSTS depends upon how fast and accurate it transfers power to the load and the cost. If GTO is deployed the overall cost will be higher due to the complexity of the controller circuit to turn on and turn off (applying current of reverse polarity) whereas the thyristor based switches are easily available at less cost. The main drawback of thyristor switch based SSTS is that unless forced commutation circuit is not used, thyristors can be turned off at natural zero crossing only. Since a low cost hardware prototype is modelled here, thyristors are used. As the system voltage is 230V, thyristors of more than three times the rated voltage need to be used. For the implementation, a back to back connected thyristor IC, SKKT27B16E as shown in Fig.6.25 with 1700V, 50A rating is used.

Fig.6.25. Two thyristor IC [18]

It has a latching current I_L=400 mA, maximum holding current I_H=200mA, gate trigger voltage V_{GT} = 3V, gate trigger current I_{GT}=150mA, dV/dt =1000V/μsec.

6.5.2.2. Development of Hybrid SSTS:

If we operate the thyristor switches for longer duration, the cost of cooling will increase due to excessive power loss. In addition to this, it can introduce voltage harmonics in the system. This drawback can be rectified using a hybrid SSTS. In this Scheme (hybrid SSTS), high speed mechanical switches are used for normal system operating conditions. During normal operating conditions mechanical switch of preferred source is closed and that of the alternate source is open. When the transfer is requested, the thyrisors of preferred feeder is switched ON and simultaneously mechanical switch is switched OFF. The current is thus commutated to thyristors of preferred feeder and blocked at next zero crossing of the current. As soon as the current is completely blocked the thyristors of alternate source is switched ON and powers the load. Once the currents through alternate thyristors stabilize, the mechanical switch of the alternate source is enabled and blocks the pulses to thyristors. In the hardware prototype, hybrid SSTS shown in Fig.6.26 is implemented.

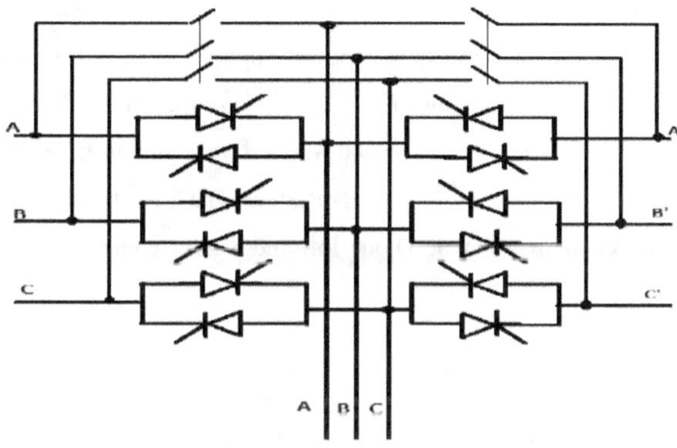

Fig.6.26. Schematic representation of hybrid SSTS

In order to model high speed mechanical switch, relays/contactors are employed for low cost and low power application which can switch within 5-20 ms.

Fig.6.27. Relay Switch [18]

The relays used are KT 240 high power relays with 12V, 40 A rating. The circuit is configured in such a way that during normal operation these relays will be short circuited and thyristors will come into operation only when switching happens.

6.5.2.3. *Firing Control of Thyristors:*

The thyristors has to be turned on at exact zero crossing of the voltage so as to prevent the harmonics that can be generated into the system. There are different methods existing for zero crossing detection, but the problem is that switching should occur at the exact zero crossing and if digital controllers are used, system voltage detection and firing will not succeed unless we use high speed processors. So analog circuits will be easy to implement. Also these circuits should be implemented with controlled a zero-crossing-firing by the control centre such that any time it can be switched off. A special IC, MOC3021 is selected which is a photo triac driver suitably modified for thyristor firing with an NPN transistor for this application. The logic diagram of the IC is shown in Fig.6.28. The IC is a Gallium-Arsenide-Diode Infrared Source and optically coupled silicon triac driver having high isolation of 7500 V.

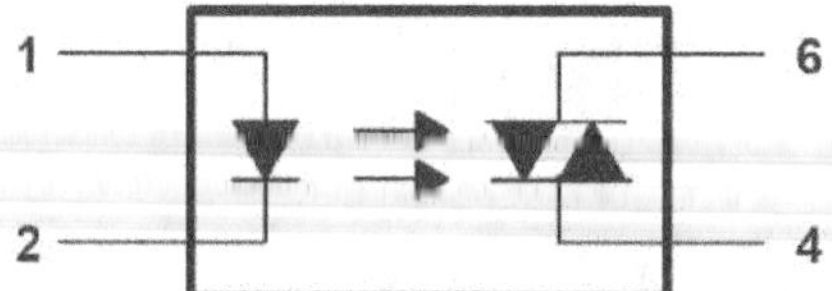

Fig.6.28.MOC 3021 IC logical diagram

The input point of the MOC 3021 is connected to the VCC to forward bias the diode and to control the switching, high speed switching transistor 2N2222A is tied to the cathode point of the diode. The control signal from microcontroller is connected to the base of the switching transistor with a resistor to limit the base current. Thus through the signals from microcontroller, the driver can be controlled to switch at any instant. The input from the step down transformer is given to the triac through a diode to limit the conduction to a half cycle. Resistors are provided to limit the current. The voltage across the resistance R is given to the thyristor gate circuit. When there is a current flow through the diode, the triac gets activated. The complete zero crossing firing scheme is shown in Fig.6.29.

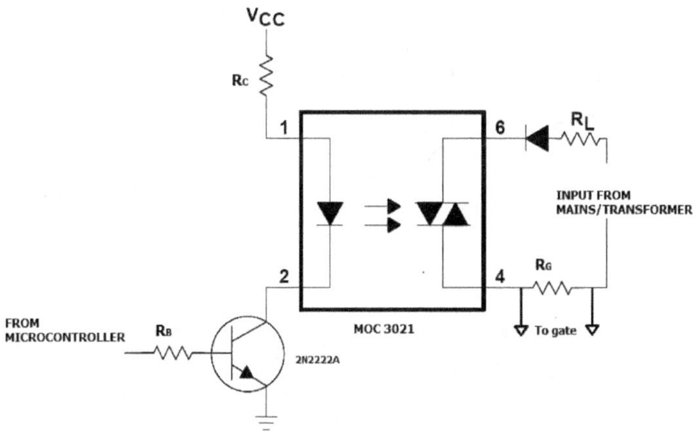

Fig.6.29. Zero crossing firing control using MOC 3021 IC

6.5.2.4. *Testing of SSTS:*

The thyristor firing is verified using the MOC3021 IC for antiparallel switch and the complete SSTS single phase model. A lower voltage of 18V rms is applied through a step down transformer to the thyristor switches and also to the firing controller. At lower operating voltages the firing control scheme will not work well as the thyristor voltage drop of around 3-6V will happen. So it is worth to use hybrid SSTS model. The operation of hybrid SSTS during switch over is shown in Fig.6.29. Initially preferred feeder (PF) was connected to load. So under normal operating condition the relays in PF are on. During switch over from PF to alternate feeder (AF), the thyristor units in the PF is switched ON and then switched OFF. Following this, thyristor units in AF is switched ON and the relay in AF is switched ON. Once the current settles to normal value, thyristor units will be switched OFF.This way switchover is done and system comes back to normal operation. The drop in the thyristors are not significant when system operates at high voltages. In Fig.6.30, there is a small time delay in which both thyristors are off, that is just to ensure the shorting of two feeders are not happening. With proper protection scheme even 'make before break' connection can be employed. However, the thyristors are on only during switching operations and with the help of that smooth transition from one feeder to another feeder happens.

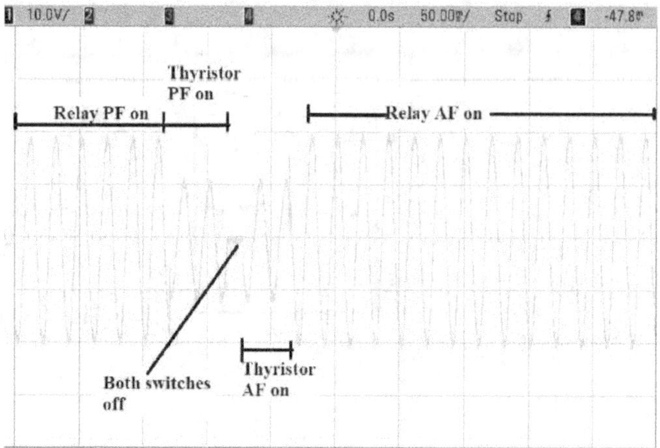

Fig.6.30. Test results with hybrid SSTS

Fig.6.31 shows the hardware circuit for hybrid SSTS implementation

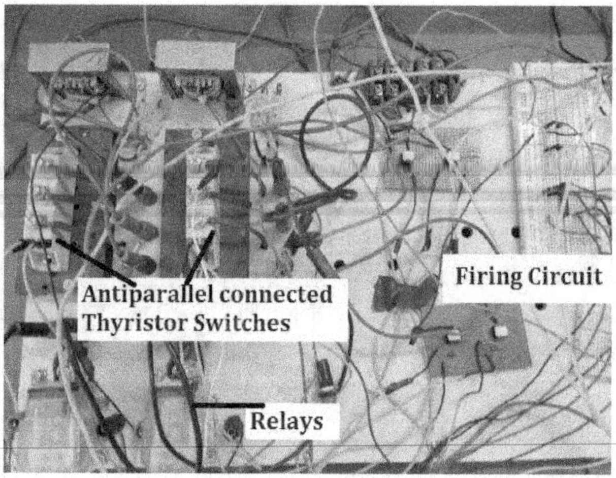

Fig.6. 31. Hardware circuit of hybrid SSTS

The hardware circuit mainly consist hybrid SSTS model composed of two thyristor IC, relay switch, driver circuit with MOC 3021 IC etc. Two sets of these devices are used corresponding to preferred and alternate feeders. Different cases considered for testing are listed below.

a) Switching ON of preferred feeder:

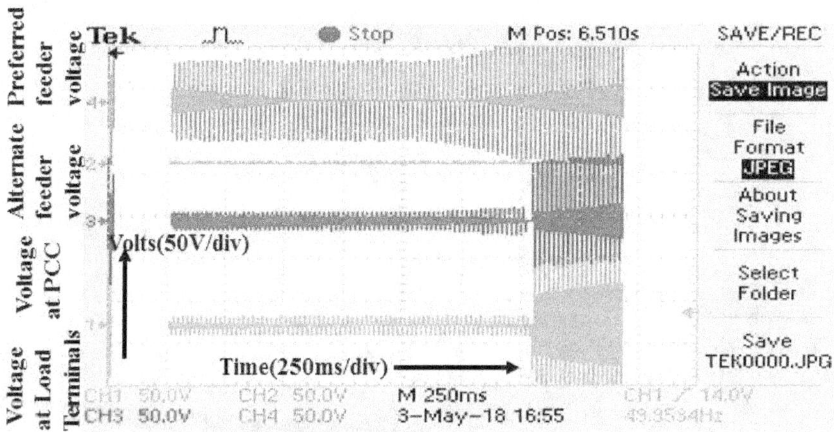

Fig.6.32. Switching on of preferred feeder

The overall controller continuously monitors the voltage at preferred and alternate feeders. Once it is above 60%, the corresponding feeder can be switched ON. The waveforms associated with switching ON of the preferred feeder is shown in Fig.6.32. The first two waveforms represent the preferred and alternate feeder voltages. Voltage at PCC and voltage at load terminals are represented as last two waveforms. From Fig.6.32, it is seen that when the voltage at preferred feeder reaches more than 60% of rated voltage, the SSTS connected in that feeder is switched ON and voltage is made available at PCC and load terminals. Since performance of SSTS needs to be tested, the DG is kept in OFF mode to see the voltage level at PCC at the instant of switching on of the preferred feeder.

b) Switching ON of alternate feeder:

The testing of SSTS during the switching ON of the alternate feeder has been carried out and the associated waveforms are shown in Fig.6.33. In this scenario, when the alternate feeder voltage is greater than 60%, corresponding SSTS in the alternate feeder is switched on and is connected to the PCC in order to make the voltage available at load terminals. To test the SSSTS performance, corresponding preferred feeder voltage is made to zero and DG is considered to be in OFF condition.

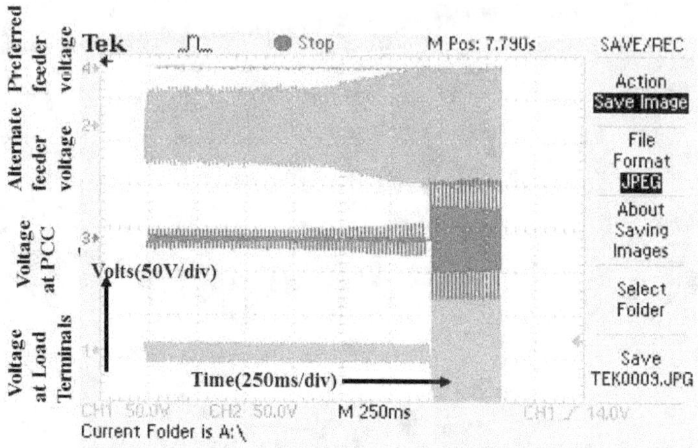

Fig.6.33. Switching on of alternate feeder

c) *Switch over from preferred to alternate feeder and vice versa:*

The main function of SSTS is to protect the loads from voltage variations which arise mainly due to faults. At the time of switching on, the pulses are given to antiparallel thyristors to trigger them. Once the current through the circuit reaches normal value, mechanical switch will be turned ON and thyristor will be switched OFF. In a similar way, when there is a switch over from preferred feeder to alternate feeder, the thyristors of preferred feeder will be turned ON first (relay will be OFF), then thyristors of preferred feeder will be OFF, following which thyristors of alternate feeder will be turned ON. Then the relays of alternate feeder will be ON, thereafter switching OFF the thyristors of alternate feeder. This way it ensures smooth transfer from one feeder to other feeder within less than a cycle time. Fig.6.34 shows the waveforms obtained with the hardware implementation of SSTS. The moment the voltage on one feeder is greater than other, switchover also happens with in less than a cycle which is clearly depicted in the Fig.6.33. The switchover time can be further reduced with the replacement of thyristors with GTO switches. The alternate feeder to preferred feeder switch over is depicted in Fig.6.35.

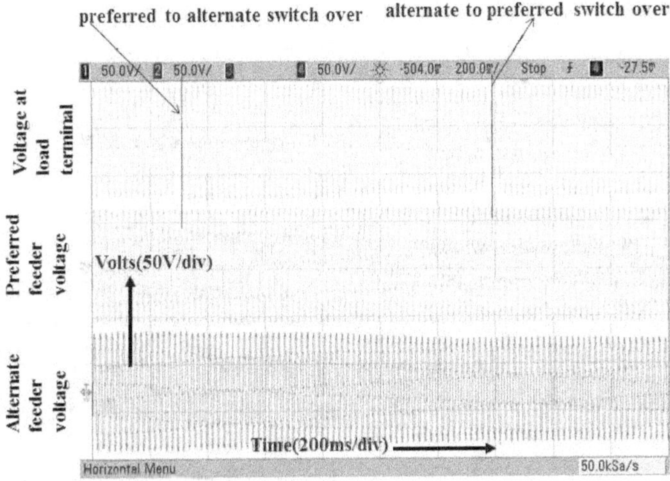

Fig.6. 34. Switchover between feeders

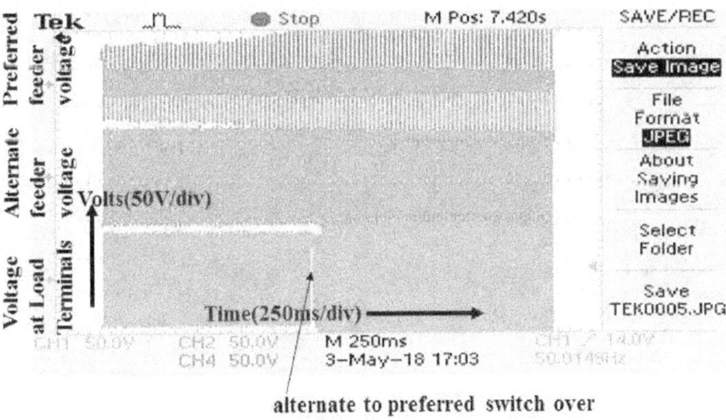

Fig.6.35. Switchover from alternate feeder to preferred feeder

Fig.6.35 represents the preferred feeder voltage, alternate feeder voltage and voltage at load terminals respectively. The alternate feeder voltage is kept constant and preferred feeder voltage is increased. When the voltage becomes greater than alternate feeder voltage, at that instant itself, SSTS connects the loads to preferred feeder.

The sensing, controlling and operating the switches are happening within less than half cycle which clearly depicts effectiveness of the sensing module, overall controller and the SSTS of the hardware setup.

6.5.3. Development and testing of Coordinated Controller for Custom Power Park

The controller for the custom power park must be implemented with high speed digital processors which can monitor the system events and switch between feeders and diesel generators to support the loads without any power failure for various disturbances which can occur in the system composed of loads and feeders. It should be capable of detecting sags, unbalances in the phases and other perturbations with high speed computational algorithms. For miniature model development it is sufficient to use microcontrollers having good processing speed and more ports to control independent breakers and devices.

6.5.3.1. Selection of controller:

The overall controller is developed using Arduino micro controller. Arduino is a platform for sensing inputs, processing them and relaying the same to a desktop computer. It is an open source physical computing platform and uses a simple microcontroller and IDE for burning software code into the board. The Arduino Uno shown in Fig.6.36, R3 is a microcontroller board based on the Atmega328. It has 14 digital input/output pins (of which 6 can be used as PWM outputs), 6 analog inputs, a 16 MHz crystal oscillator, a USB connection, a power jack, an ICSP header, and a reset button. It contains everything needed to support the microcontroller It simply connect it to a computer with a USB cable or power it with an AC-to-DC adapter or battery to get started [18].

Fig.6.36. Arduino Uno board[18]

Arduino has the advantages like low cost, cross platform, simple and easy programming environment, open source, extensible software and hardware environment etc.

The specifications of Arduino board are shown in Table 6.5.

Table 6.5: Specifications of Arduino

Microcontroller	ATmega328
Operating Voltage	5V
Input Voltage (recommended)	7-12V
Input Voltage (limits)	6-20V
Digital I/O Pins	14 (6 provides PWM output)
Analog Input Pins	6
Dc Current per I/O Pin	40mA
Dc Current for 3.3V Pin	50mA
Flash Memory	32KB(ATmega328)
SRAM	2KB(ATmega328)
EEPROM	1KB(ATmega328)

Arduino IDE is a cross platform application written in Java and is derived from IDE for Processing programming language and wiring projects [18]. It is designed to introduce programming to beginners who are unfamiliar with software development. It includes a code editor with features such as syntax highlighting, brace matching, automatic indentation and also can compile and upload programs on to the board with a single click. A program written in Arduino IDE is called a sketch. It comes with a C/C++ library called wiring which makes common input/output operations easier. Programs are written in C/C++, and users need only define two functions to create a runnable program.

i.   Setup ()- A function that runs once at the start of the program that can initialize settings.

ii.   loop ()- A function that runs repeatedly until board powers down

6.5.3.2. Interfacing Coordinated Controller with sensors:

The feeder voltages are connected to step down transformer of 230/12 V rating and then to DF10 series single phase rectifier IC with capacitive filter to get pure DC. The output of the rectifier IC is connected to a simple DC voltage sensor of 25V rating. This has 5 pins as seen in Fig.6.37. The S pin is connected to the analog input of Arduino. The positive pin is not connected to anything. The negative pin is connected the ground of Arduino. The GND pin in the voltage sensor is connected to the lower side of the voltage whose measurement needs to be taken and the VCC pin is connected the higher side of the voltage. The voltage sensor is integrated with Arduino Uno board which reads the sensor output and sends it to the master controller.

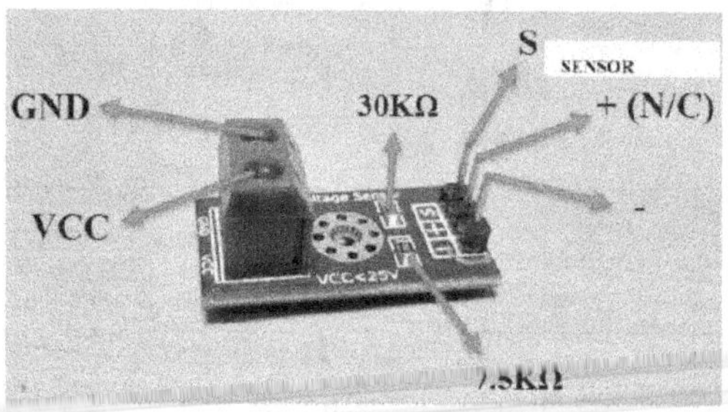

Fig.6.37. DC voltage sensor [18]

6.5.3.3. Interfacing Coordinated Controller with SSTS:

The Arduino controller produces output based on the sensed input voltages. Coding has been done for the control logic and based on that the signal corresponding to the SSTS connected to preferred or alternate feeder can be activated. These 5V control output pulse from output port of the microcontroller is connected to the base of the transistor shown in Fig.6.29. Then MOC3021 IC gets activated, which in turn triggers the thyristor switches in SSTS.

6.5.3.4. Interfacing Coordinated Controller with Breakers and Relays:

The relays/ contactor / breakers etc. need high current ranging from hundreds of milli amperes to few amperes and 12-230V AC/DC voltage for energizing the control coils which is not possible with a microcontroller 5V, few milli amperes current. So proper interfacing is required to control different breakers through a 5V control pulse. To control 12V-50V relays/contactors relay driver IC ULN2803 is required. The ULN2803 device is a high-voltage, high-current Darlington transistor array. The device consists of eight NPN Darlington pairs that feature high-voltage outputs with common-cathode clamp diodes for switching inductive loads. The collector-current rating of each Darlington pair is 500 mA. The Darlington pairs may be connected in parallel for higher current capability. The device has a 2.7kΩ series base resistor for each Darlington pair for operation directly with TTL or 5V CMOS devices. For interfacing with 230V single phase control coils of contactors or large breaker ratings, the same IC can be made use of along with small relay of operating voltage of 12-50 V DC. It can be connected to 8 separate relays and can be operated independently by giving control pulse.

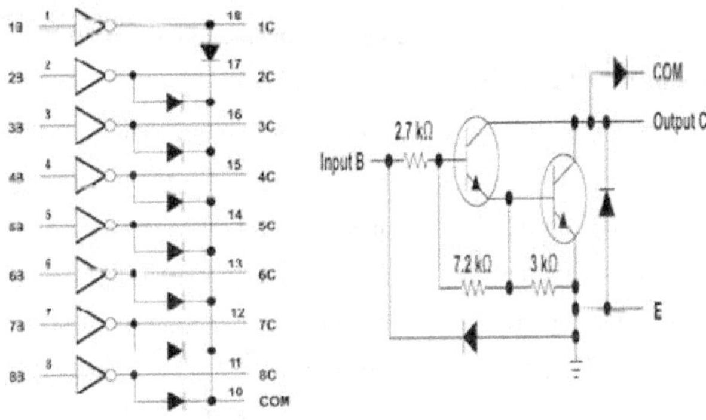

Fig.6.38. Logical diagram and basic schematic of each Darlington pair [18]

The relays /breakers are connected to each Grade A, Grade AA and Grade AAA categories of load, Shunt and Series Active Filters, DG, common bus, SSTS in preferred and alternate feeders etc. The control signal from output port of microcontroller is connected to relay driver IC which in turn drives the control coil of the relays. The relay setup along with various connection points are shown in Fig.6.39. It has one common bus connecting various static and mechanical switches and is also connected to the CPP common bus point, to a diesel generator, to active filters and to the AA, AAA load points. The incoming point is connected to the common breaker and fed to another bus for power conditioning. It is then connected with the custom power park controller through relay driver IC for controlled operation.

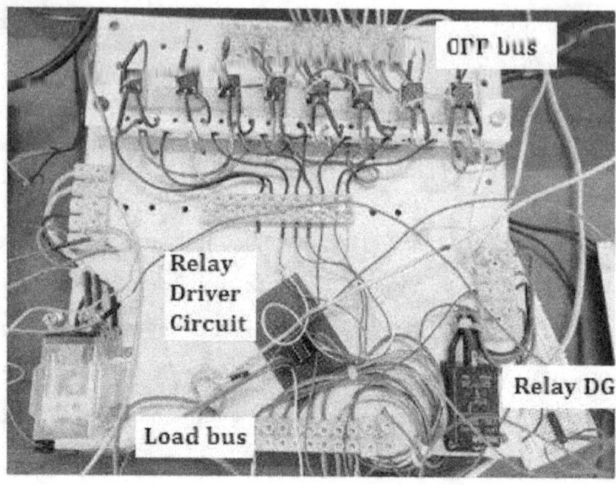

Fig.6.39. Relay Setup

For hardware implementation, voltage sensors are used to sense the voltage at both feeders. The sensed voltage is given to a microcontroller which acts as the main controller. Based on the voltages at the feeders, switching decision is taken. The coding is done based on the flow chart shown in Fig.6.2.

Fig.6.40 depicts the action of the coordinated controller. It represents Load AAA, Load A, Load AA and preferred and alternate feeder voltages. A high sag has occurred in both preferred and alternate feeders simultaneously. At the instant of threshold value of the voltage level, the controller isolates the preferred and alternate feeders from the loads and before DG starts, the AAA loads are only powered by the series active filter with proper BESS backup. As the DG get started after the delay, AA and AAA loads are powered by means of DG while the A loads are not powered until any one of the feeders' voltage is restored as in Fig.6.40. All these actions are performed by the coordinated controller.

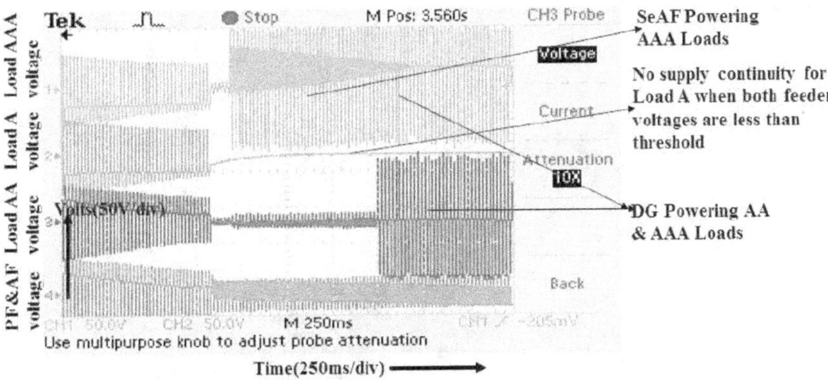

Fig.6.40. Custom Power Park AAA loads powered with series active filter feeder, the A and AA loads are not powered before DG start and after DG start AA and AAA are powered by DG

Fig.6.41 represents Load AAA, A, AA voltages along with preferred and alternate feeder voltages during the starting or synchronization time delay. Load AAA gets full voltage at its terminals as Series Active Filter feeds power to it, whereas the continuity of the supply is lost for Load A and Load AA during DG start up time.

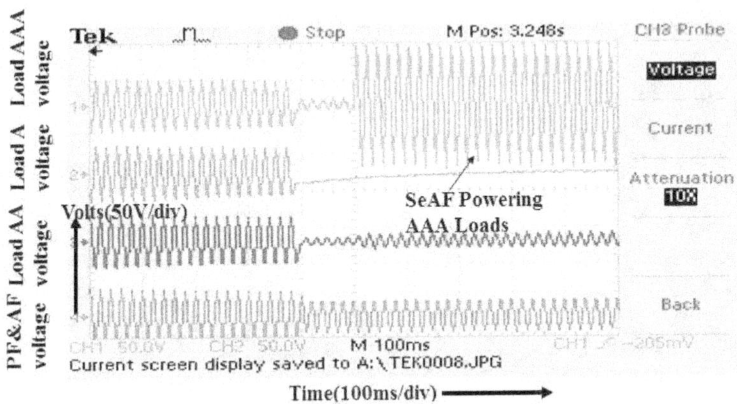

Fig.6.41. Custom Power Park AAA loads powered with series active filter while the A and AA loads are not powered before DG start

When both feeder voltages are less than 0.6pu, then DG has to be switched on. Once the DG ready is signal is high (indicates DG is ready to feed the load), then the overall controller turns on the switches so that the DG can be connected to feed AAA and AA load which is clearly depicted in Fig.6.42.

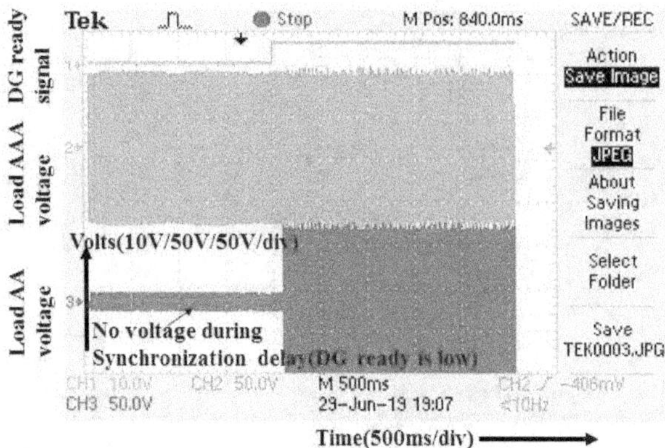

Fig.6.42.DG ready signal, voltage at AAA load, voltage at AA load

The status of different switches during switching ON of preferred feeder is shown in Fig.6.43. Since both the feeder voltages are less than 0.6pu, DG is ON. In that case SSTS in both feeders and Load A breaker is OFF. When preferred feeder voltage reaches 0.6pu, preferred feeder SSTS is ON and the message, 'Vp started', is displayed on the screen. At this time SSTS in alternate feeder is still OFF and breaker connected to load A is turned ON resuming the services. Breaker to DG is turned OFF because preferred feeder is connected to CPP bus and then to loads.

```
COM5 (Arduino/Genuino Uno)

volt_1: 0.58,volt_2: 0.00
 DG is already on
Ctrl STS_p = 0,Ctrl STS_a = 0,Common Bus = 0,Brk A = 0,Brk AA = 1,Brk AAA = 1,DG = 1,Shunt AF = 1 Series AF = 1
volt_1: 0.81,volt_2: 0.00
Ctrl STS_p = 1,Ctrl STS_a = 0
vp started
following setup is going to be made
Common Bus = 1,Brk A = 1,Brk AA = 1,Brk AAA = 1,DG = 0,Shunt AF = 1,Series AF = 1,DG start = 0
volt_1: 0.64,volt_2: 0.00
Ctrl STS_p = 1,Ctrl STS_a = 0
preffered voltage running
Common Bus = 1,Brk A = 1,Brk AA = 1,Brk AAA = 1,DG = 0,Shunt AF = 1,Series AF = 1,DG start = 0
volt_1: 0.65,volt_2: 0.00
Ctrl STS_p = 1,Ctrl STS_a = 0
preffered voltage running
```

Fig.6.43. Status of different switches when preferred feeder is switching on

The status of different switches during switch over from preferred to alternate feeders using SSTS is shown in Fig.6.44. In this case preferred feeder is already ON. When the alternate feeder voltage is just greater than that of preferred feeder voltage, switchover between feeders happens and after that it displays the message that alternate feeder is running. The SSTS corresponding to preferred feeder is turned OFF and SSTS corresponding to alternate feeder is turned ON thereby connecting that feeder to CPP bus.

```
COM5 (Arduino/Genuino Uno)

volt_1: 0.65,volt_2: 0.64
Ctrl STS_p = 1,Ctrl STS_a = 0
preffered voltage running
Common Bus = 1,Brk A = 1,Brk AA = 1,Brk AAA = 1,DG = 0,Shunt AF = 1,Series AF = 1,DG start = 0
volt_1: 0.65,volt_2: 0.65
Ctrl STS_p = 0,Ctrl STS_a = 1
stwitched to alternate from preffered voltage
Common Bus = 1,Brk A = 1,Brk AA = 1,Brk AAA = 1,DG = 0,Shunt AF = 1,Series AF = 1,DG start = 0
volt_1: 0.65,volt_2: 0.66
Ctrl STS_p = 0,Ctrl STS_a = 1
alternate voltage is running
```

Fig.6.44. Status of different switches during switchover from preferred to alternate feeder

The status of different switches during turning ON of DG is shown in Fig.6.45. When both feeder voltages are below 0.6pu, then DG is tuned on and DG breaker status is shown as 1. STS breakers to both feeders will be OFF.

```
COM5 (Arduino/Genuino Uno)

volt_1: 0.60,volt_2: 0.47
DG started on
Ctrl STS_p = 0,Ctrl STS_a = 0,Common Bus = 0,Brk A = 0,Brk AA = 1,Brk AAA = 1,DG = 1,Shunt AF = 0,Series AF = 1
volt_1: 0.59,volt_2: 0.47
 DG  is already on
Ctrl STS_p = 0,Ctrl STS_a = 0,Common Bus = 0,Brk A = 0,Brk AA = 1,Brk AAA = 1,DG = 1,Shunt AF = 0,Series AF = 1
```

Fig.6.45. Status of different switches during switching on of DG

For the results which are demonstrated through Fig.6.31 to Fig. 6.34, the necessary control signal was sent by overall coordinated controller. Thus in the custom power park, the coordination of various network reconfigurable and compensating type devices and also different loads has been taken care by the coordinated control scheme. This is the crucial part in the implementation of CPP and even a small mistake can give rise to mal-operation of so many devices in the park. The proposed overall controller has taken utmost care about this.

6.5.4. Development and testing of FVPD controller for SeAF

The most crucial element in CPP is the series active filter which maintains a distortion free voltage with rated magnitude at the sensitive load terminals thereby protecting them from any voltage fluctuation. A novel controller, namely the Fundamental Voltage Peak Detection Controller [explained in chapter 5] is used to generate the reference voltage signal for the Series Active Filter.

The series Active filter injects voltage at terminals which gets added with PCC voltage to maintain load terminal voltage at rated value. The analog implementation of novel controller is done and tested under different conditions of voltage disturbance. A diode bridge rectifier of 300W is used as nonlinear load for the hardware implementation.The voltages are sensed by EL100P2 voltage sensor and given to three phase analog controller made up of mainly 741 ICs except muliplier and divider ICs.The reference filter voltage is added with the sensed voltage to get the load voltage after compensation.In Fig.6.46, different cases like harmonics, sag, swell and unbalance are considered.In each case the reference voltage waveforms are shown in the second row.The load voltage becomes sinusoidal with rated magnitude in all the above voltage related issues.

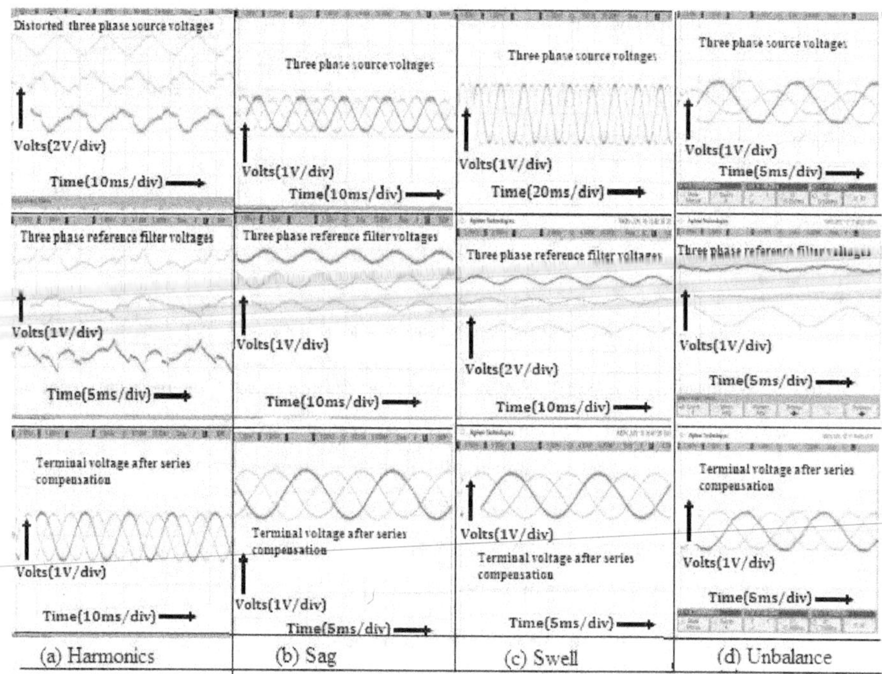

Fig.6.46.Source Voltage before compensation,Reference Filter voltage and voltage at laod terminal after compensation

6.5.5. Development and testing of Communication Module

The proposed solution involves the design and development of a controller for the power quality improvement. The controller needs to collect the voltage levels from different available feeders and compare them and make decision. The collected information is passed to the central control unit. The control unit recieves the data from feeders and make a decision based on the logic developed. The decision needs to be communicated to various units such as loads, compensating devices, SSTS etc. This makes the whole setup as a three module system with communication facility between these modules. The modules are sensing units at feeders, main control unit with decision making part and loads, compensating devices etc. with communication modules.

The block diagram of the proposed system is provided in Fig. 6.47. The three major parts are Sensing Unit at the feeders, Master Control Unit and Loads with Communication modules. Once the modules and its functionalities are identified, the next task is to establish communication between them. The Sensing Units at feeders need to pass the data to the Master Control Unit and no receiver functionalitiy is required. Similarly, the Loads with communications modules need to receive control information from Master Control Unit and no need to send any data back. The master control unit need to receive data from sensing units and send control instructions to loads, compensating devices and SSTS. Considering all these communication requirements and the fact that the different modules can be located at different locations, it is recommended to have wireless communication rather than wired communication. This decision is also based on the various observations about the context in which the proposed system is going to function. Wired medium will face more noise due to the high voltage transmission lines and also the fact that the feeders and Master Control Unit can be mobile in nature are the factors which propel the decision to go with wireless medium.

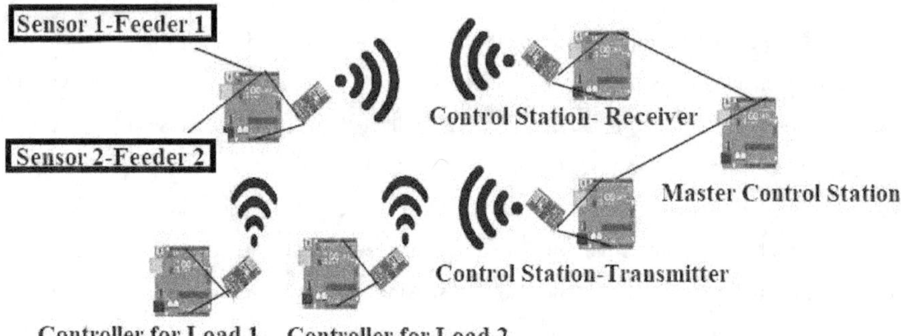

Fig.6.47.Block diagram of communication system

Even though wired media is simple and highly reliable in general conditions, in the proposed system, wired media will suffer from high interference. Due to this, it is decided to go with wireless media. ZigBee, WiFi, Bluetooth etc. are various choices for wireless media. Bluetooth and ZigBee are short range communication models and WiFi is flexible and effective when long range communication is required. So it is decided to conduct the experiments with both ZigBee and Wifi communication models.

In the proposed system, Arduino boards are used as microcontroller units for either interfacing the sensors or building the main controller. Node MCU and ESP8266 are two different options for providing WiFi services with arduino. Out of these, ESP8266 is simple and low cost. ESP8266 is hence chosen to provide wifi communication. Similarly, XBee modules are used as alternative wireless medium of communication.

<u>Sensing Units at the Feeders:</u>

The voltage levels at the feeders need to be sensed. Voltage sensors are used to sense the voltage levels at the feeder. These sensed data needs to be communicated to the Master Control Unit. The Sensing Unit is setup using Arduino boards placed near the feeders which are equipped with voltage sensors. These voltage sensors collect the voltage level and pass it to the Master Control Unit using the communication module attached. The Sensing Unit is equipped with ESP8266 WiFi module to send the data to Master Control Unit.

<u>Master Control Unit</u>:

The Master Control Unit is responsible for collecting the feeder data and make a decision on the status of loads and other devices. This control unit is again a microcontroller unit with ESP8266 WiFi module / XBee module to send and receive data from other units. Once the data is received from feeder units, the controller works based on the control algorithm loaded in the main controller. The control algorithm makes the decision of switch over between feeders or switching on the DG and pass this control instruction to the WiFi module to pass it to the corresponding control station.

<u>Control Stations for Loads</u>:

The control station is also a microcontroller unit with ESP8266 WiFi module / XBee module which will act based on the control instruction it has received from the Master Control Unit via wifi module.

6.6. Conclusion

In this chapter, the hardware implementation of the prototype of a Custom Power Park (CPP) is discussed, which is a dedicated power entity which buys power from grid and employs different Custom Power Devices (CPDs) to provide clean, high quality, steady and highly reliable power to its end users according to their requirements of quality of power service, at higher prices than grid prices. The Custom Power Park employs Custom Power Devices and a Coordinated Controller which monitors, makes decisions, controls and coordinates the network. Voltages in preferred and alternate feeders are sensed and data is sent to a central controller. Based on the data received from sensors and from customer requests, the central controller decides what all custom power devices need to be switched ON, whether switchover between preferred and alternate feeders using a Solid-State Transfer Switch (SSTS) is required or not and sends appropriate control signals to respective devices in the park thereby making the system fully autonomous. The performance of the Coordinated Controller under various system conditions and the functioning of the Solid-State Transfer Switch (SSTS) and other CPDs are analyzed in this chapter. The development of CPP also involves the financial aspects such as operational cost, cost for buying power and cost per unit of the power provided. But these aspects are not discussed as these decisions are beyond the scope of this work as in Indian context, these decisions are managed by the government agency which deploys the CPP.